978-0151123629
AF521877

Biology and Man

Books by George Gaylord Simpson

ATTENDING MARVELS
1934

TEMPO AND MODE IN EVOLUTION
1944

THE MEANING OF EVOLUTION
1949

HORSES
1951

LIFE OF THE PAST
1953

THE MAJOR FEATURES OF EVOLUTION
1953

LIFE: AN INTRODUCTION TO BIOLOGY
With C. S. Pittendrigh and L. H. Tiffany
1957

LIFE: AN INTRODUCTION TO BIOLOGY
Second edition. With William S. Beck
1965

QUANTITATIVE ZOOLOGY
Revised edition. With Anne Roe and Richard Lewontin
1960

PRINCIPLES OF ANIMAL TAXONOMY
1961

THIS VIEW OF LIFE
1964

George Gaylord Simpson

Biology and Man

Harcourt, Brace & World, Inc., New York

Copyright © 1964, 1966, 1967, 1969 by George Gaylord Simpson

Chapter Ten is copyright © 1966 by the American Psychological Association, and is reproduced by permission (see page 164).

All rights reserved. No part of this publication may be reproduced or transmitted in any form or by any means, electronic or mechanical, including photocopy, recording, or any information storage and retrieval system, without permission in writing from the publisher.
First Edition
Library of Congress Catalog Card Number: 69-12045
Printed in the United States of America

Preface

Biology and man (meaning all of mankind) are the two most interesting subjects in the world. When considered concurrently, there are several possible approaches to them. First, biology can be treated as a human activity. It, and no longer mathematics, is now the queen of the sciences. How far it has come, where it stands today, what it may become—these are of extraordinary importance among questions about man's intellectual attainments and prospects. Second, knowledge of biology and the applications of that knowledge are affecting the lives of all of us. Everyone should have some concern with that fact, not only to understand our present situation but also to look to its obviously needed amelioration. Third, man himself can be considered as a biological phenomenon and a subject for biological study. Certainly this is not the only way to study man, nor does it reveal the whole man, but just as certainly it is an absolutely essential part of that attempted revelation.

Those are the main themes of this book, and they are here taken up in that order. They are not sharply bounded. One leads to the next, and the chapters are not assembled in parts. There is some partial repetition of included subjects as the point of view shifts from that of biology as a science to that of man as a biological species.

The problems of biology in the modern world are not just those of obtaining and explaining more facts about organisms. They are also organizational, and these are problems that stand out when one considers biology not only as a field of knowledge and research but also as an activity pursued by human beings. In this respect it is no secret that certain stresses and strains develop from time to time. In recent years some of these have resulted from the fact that one of the older branches of biology, now often going under the lately coined term "molecular biology," has recently made a spectacular breakthrough. It is important to consider the nature of that breakthrough, its relationships with other branches of biology, and the need and possibility for exploiting it more fully in that broader context. Those are subjects of the first three chapters of this work.

Among the many effects of biology on our lives, the medical applications are most obvious and most generally known. In Chapters Four and Five I have therefore taken those aspects for granted and have paid attention rather to the involvement of biology in our philosophies and outlooks, in the drive for bettering our now deteriorating ecological circumstances, and in our culture as a whole, which is changing in various ways, some for better and some for worse.

Looking at man as a biological species, some biologists, professional and amateur, have become so preoccupied with the fact that man is an animal that they have neglected the fact that he is an absolutely unique animal. It would be difficult to say which fact is more important. Fortunately, the decision is unnecessary, and indeed it should not be made. Discussion of man's animality without reference to his humanity or of his humanity without reference to his animality is sterile to the point of inanity. Some lately popular books all too amply illustrate the folly of both extremes. I have tried to avoid those follies in Chapters Six to Eight, which treat of those matters.

A point essential in consideration of both man's animality and his humanity is that he originated by evolution. Both his organic evolution and his cultural evolution, a form of biological adaptation, will continue. It is tempting but hazardous to consider where those evolutions may lead in the future, and I have yielded to that temptation in Chapter Nine, with a moderation that may seem timid to some. Finally, it goes beyond the scope of biology as a science to treat of ethics, but this can be done from a biological point of view, as it is in Chapter Ten. This is, after all, one of the most important ways in which biology can and I believe should affect our lives.

Seven of ten chapters have appeared in serial publications elsewhere. The circumstances of original composition and first publication are given in the appended section, "Sources and Second

Thoughts," along with a few sources in the other sense of authorities for some statements made and a few suggestions for further reading. Chapters previously published as separate essays have not been materially modified. Although all have been written rather recently (the oldest in 1964), in some instances I have had second thoughts, either as to modification or as to amplification. Those, too, are appended.

It is obvious that the views here expressed are those of the author, and I hope it is untrue that they are not the views of anyone else. No one else does share responsibility, but I want to express a continuing general indebtedness first, and always, to Dr. Anne Roe, my wife, and then to all the appreciated colleagues at Harvard University and at the University of Arizona. Thanks go also to Kitty Lance, who prepared the manuscript.

G.G.S.

Tucson, Arizona
May 1968

Contents

Biology and Man

CHAPTER ONE

The Crisis in Biology

"Each pursues his own theory, little solicitous to correct or improve it by an attention to what is advanced by his opponents."
—T. R. MALTHUS (in 1798)

A few years ago I overheard two clever young instructors talking about courses they were giving in genetics. One said, "I don't mention genes until the last part of the course." The other rather scornfully said, "I never mention genes at all." Around that time one of their elders was developing and publicizing a new course in general biology in which, as he boasted, organisms were not talked about until the second term, and then only as if they were the test tubes in which chemical reactions take place.

When bright young biologists speak of genetics without genes and wise old biologists of life without organisms, it is evident that something peculiar is going on in the science of biology, so peculiar that "crisis" is not too strong a word. I would diagnose this as combining monomania and schizophrenia. These ailments are rarely fatal, but they are seriously disabling. Fortunately for biology, recovery can be predicted. In fact, I think that recovery is now under way. The cited instructors and professor, so anxious to be in the scientific *avant-garde*, are already out of date unless, like many of their colleagues, they now have changed their minds.

What those teachers were reflecting and what led to the crisis was, of course, the splitting of biology into physical, chemical or molecular biology on one hand and organismal or evolutionary biology on the other. Even so, that would not have been more pathological than any other of the inevitable specializations in science if it had not somehow inspired a missionary fervor for molecular biology as *real* biology, the biology of the future, and so on, along with scorn in some quarters for organismal biology as passé and currently sterile. Those attitudes were intensified by recent successes in studies of nucleic acids—a point that becomes almost comically ironic when one recalls that it was the organismal, not the molecular, biologists who discovered the roles that the nucleic acids play in organisms, their great importance in all living systems.

Although the split in biology eventually came to center on the nucleic acids, the various DNA's and RNA's, or to be symbolized by them, it had origins in attitudes and philosophies long before those molecules were even known to exist. It is not true that molecular biology is a new science, which has taken over (or could conceivably take over) where evolutionary or organismal biology leaves off. (For that matter, of course, it is not true that evolutionary biology *has* left off!) The oxygen cycle in both animals and plants was already well known in the eighteenth century, and the origins of molecular biology could be traced even further back. It had already long existed as a specialty when it entered a new era by Wöhler's synthesis of urea in 1828. Evolutionary biology hardly existed before 1859. Thus molecular biology is, in fact, older than evolutionary biology, although it is true that some organismal aspects of biology antedated them both. One would expect that the older molecular biology and the newer evolutionary biology would advance hand in hand once the latter arose. The fact that they did not do so seems downright irrational in retrospect. It is this that leads to the partial diagnosis of biology's ailment as schizophrenia.

I have said that monomania was also involved, and this comes in on a philosophical basis. Science, strictly speaking and in the modern sense, began its development only about four centuries ago. It was not an outcome of such earlier vagaries as astrology or alchemy, or even of the earlier useful arts or technologies, although at first it borrowed some concepts and techniques from them. It was largely a reaction against pseudosciences, antisciences, subjective philosophies and theological authoritarianisms. Haltingly at first, but finally decisively, it achieved an objective approach to the study of the universe. It developed the formation of hypotheses by induction from repeatable observations and the testing of those hypotheses by deduction of other, also repeatable observations. The latter observations were predicted, and if the prediction was confirmed, the conditions involved in it were taken as explaining the outcome. This attitude toward science led to the conclusion, or rather it seemed to require the assumption, that the physical universe is materialistic and mechanistic, that as we observe it, it operates by natural, invariable laws unaffected by the immaterial or the supernatural.

As this new philosophical view of the universe developed, biology was at first excluded from it. It was felt, indeed one can fairly say that it was observed, that living things are radically different from nonliving, that life involves something more than is found in nonliving matter. It clearly embodies physical phenomena, to be sure, but in some way it also transcends them. One aspect of that opinion was that organic compounds were believed to originate only in living systems. Wöhler's synthesis of urea, although in fact somewhat equivocal, made that view untenable. Now we know that organic compounds, even the most complex of them, can be synthesized outside of living organisms and do not involve any chemical principles different from those in inorganic compounds. That and subsequent related developments led the physical scientists, the biochemists, eventually the molecular biologists, and the philosophers of science to bring biology into the naturalistic world view earlier reached by the physical scientists. Most of them did so not by expanding that view to include the special phenomena of life but by assuming that those phenomena are not really special and can be reduced completely to those of the physical sciences. That attitude, dating from more than a century earlier, was expressed anachronistically in 1965 by an eminent biochemist as follows:

For biology, generally, probably the most important single conclusion reached to date is the conviction that living systems *do* indeed obey the physical and chemical laws that govern the rest of the universe, that the detailed working of the living organism *is* amenable to exploration by physical and chemical probing, and that the properties of living organisms are *totally* comprehensible in chemical terms. [Italics as in the original.]

No biologist could now object to the first two clauses of the conclusion; nowadays they are practically truisms. Only the last clause confirms the diagnosis of monomania: comprehension of living organisms is to come from chemistry and *nothing else.* Those who shared that view maintained that it is required by the nature of science, that to admit anything more than chemistry into biology is to return to biology's prescientific phase, when organic compounds were considered impossible of inorganic origin and life was viewed as a transcendental mystery not amenable to scientific analysis.

When those philosophical attitudes were forming and hardening, the principal opposing argument was that living things do exhibit phenomena that had not been reduced to a chemical or any other strictly naturalistic basis. That was believed to be especially true of the seeming purposefulness of organisms, the fact that all of them are clearly and intricately adapted to their environments and their ways of life. The *operation* of those adaptations might conceivably and as an extreme possibility be explicable in entirely naturalistic terms (even so, not exclusively chemical), but how on earth could one explain their *origins* except by some vitalistic or supernaturalistic cause?

It was Darwin who essentially solved that problem, although it took some time for his solutions to be generally accepted. He proposed a purely naturalistic process, natural selection, objectively known to occur and now (a century after Darwin) fully demonstrated to suffice as an explanation of adaptation, of apparent purposefulness, of teleology in biology. One would think that this would enable biology to enter at once into the naturalistic world picture alongside of the physical sciences, and that conclusion has been reached by almost all evolutionary biologists. Curiously enough and tragically enough, however, the incipient rift between what we now call molecular and evolutionary biologists thereafter tended more to widen than to close. Evolutionary organismal biology requires amplification of the philosophy of science to include its special features. It is not reducible to a philosophy taking account only of the physical, nonbiological aspects of the universe. The principles of evolutionary biology do not contradict anything in physical science. They do go well beyond the principles inferable from nonliving atoms and molecules, but without becoming anything other than naturalistic. The contrasts are of literally vital importance and will next be specified and briefly discussed.

First, the nonliving systems studied by physicists and chemists are simple. They involve few kinds of materials and few variables. Such systems as occur in nature are still further simplified in the laboratory by restricting the variety of materials (such as atoms or particles) and controlling variables. The very simplest living organisms (I exclude viruses from that category) are enormously more complex than any inorganic system. Organisms do in fact have characteristics and processes that do not occur in conjunction in nonorganic materials and reactions. Those who believed in a vital, nonmaterial element peculiar to organisms were quite right to this extent. But they were wrong as to the nature of the element. It is organization. It is the complexity and the kind of structural and functional assembly in living organisms that differentiate them from nonliving systems.

There is a hierarchy of complexity that runs from atoms through molecules, cells, tissues, organs, individuals, specific populations, communities and comprehensive ecological systems to the whole realm of the organic and its environments in space and time. The lowest level that has all the basic properties of life is the cell, and biology, strictly speaking, covers the levels from there onward. (Since biology is the study of life, and molecules, as such, are not alive, the term "molecular biology" is self-contradictory.) Each level of the hierarchy includes that below. Knowledge of included levels is necessary but is not sufficient for complete understanding of those more inclusive.

That works both ways. Knowledge of molecules is necessary but not sufficient for understanding of cells. Knowledge of cells is also necessary but not sufficient for understanding of the molecules in them. For instance, you cannot understand a transistor—indeed its very existence becomes incomprehensible—unless you know what function it fulfills in an integral electronic system. Intracellular chemical reactions would likewise be incomprehensible and also would not exist if it were not for their roles in the life of the cell and the organism as a whole. And that is true both up and down the hierarchy, from an atom to the whole of life and back again. It is ridiculous to base a philosophy of science or a concept of scientific explanation wholly on the nonbiological levels of the hierarchy and then to attempt to apply it to the biological levels without modification.

A second contrast is that the physical sciences are for the most part typological and idealistic. I mean by that, that they usually deal with objects and events as invariant types, not as individuals with differing characteristics. They generally define objects and events by idealized specification of what they have in common, not by a description of what each really is. For example, all atoms of a given isotope are assumed to have precisely the same properties; they are examples of a type, not individual things. The geometrical lines and curves of physical analysis are ideas, not anything existing in the real world. Light, as studied in physics, is either a wave or a particle, or both, or perhaps neither. In any event, it is a type that cannot be described even as an idea in noncontradictory terms. Representatives of types considered invariant as such, molecules in a gas, for example, are understood to vary in their states, such as direction and velocity of motion, and the applicable physical laws are taken to be statistical. That is not done in order to account for their variations, but in order as far as possible to cancel them out in the face of what is considered an indeterminacy inherent in the nature of the universe.

I do not question that such approaches and their philosophical implications are appropriate to the current state of physical science. They are therefore also appropriate to the subbiological levels of the hierarchy of organization. They are definitely *not* appropriate for phenomena special to the biological levels. Organisms are not types and do not have types. They are individuals and no two are likely ever to be exactly alike. Their diversity is not incidental, not interpretable as deviation from a norm or type. It is part of their nature as organisms, a universal and necessary phenomenon of living things at the populational (or specific) level. If organisms did not vary, and had not done so for some billions of years, they would not exist at all. Statistical methods are also widely used in modern biology, but not as any counterploy against an immanent indeterminacy that makes their properties or activities inherently statistical. In biology statistics are used chiefly in these ways: to describe and measure individual variation and general organic diversity; and to infer characteristics of variable populations from samples and define the confidence that may legitimately attach to such inferences.

A third major contrast between the physical and the biological sciences is that the physical sciences are nonhistorical for the most part. There are certain exceptions—for example, historical geology, which is usually classed among the physical sciences—but almost the entire content of physics and chemistry has no historical element. The characteristics of objects and forces studied are those considered immanent in the universe, that is, built-in and unchanging. Gravity, for example, exists regardless of past, present or future; it has no history. Similarly a chemical reaction, even though it proceeds in time, is completely devoid of a history; it always has and always will occur in the same way under the same conditions.

It is that lack of any historical factor that makes observations in the physical sciences indefinitely repeatable, that makes exact predictions possible, and that makes prediction and explanation simply different ways of looking at the same conclusions. Philosophers of science have repeatedly insisted that these principles apply to science in general, including biology to the extent that it is "really" a science. These principles do apply to biochemistry and even to molecular biology if those subjects are pursued without reference to cells, organisms or populations, but that is because biochemistry and molecular biology are not biology at all if they are pursued in so narrow a manner. Biology that is truly such, that is, a study of living things, inevitably and always has a historical factor, and the physical principles of repeatability, predictability, and parity of prediction and explanation do not apply to the historical aspects of biology.

Organisms, of course, have various characteristics in common, in degrees varying from such minimal resemblance as between, say, a man and a sequoia to the maximal resemblance of identical twins. Nevertheless, no two organisms, not even identical twins, are exactly alike. Each is the product of a history both individual and racial, and each history is different from any other, both unique and inherently unrepeatable. These aspects of biology deal not with the immanent, the inherent and changeless characteristics of the universe, but with contingency, its states, fleeting and in ceaseless change, each derived from everything that went before and conditioning everything that will follow. The possibilities of prediction are loose and limited, in principle because contingent states are unique and never exactly repeated, and even more so in practice because the historical antecedents are enormously complex and practically unknowable in complete detail. These facts also rule out the parity of prediction and explanation. Part of the explanation of what an organism is obviously depends on what its ancestors were, what changes have occurred, and why and how. This is explanation after the fact, a posteriori, or by what has been called postdiction. It is quite different from prediction, and the possibilities of prediction in an evolutionary sequence are decidedly limited. Thus principles firmly advanced as applicable to the philosophy of science in general are not in fact applicable to some of the most important aspects of biology. In this crisis surely it is obvious that the solution is not to restrict the scope of biology, practically excluding its most characteristic and most important aspects, but to broaden the scope of scientific principle and philosophy.

That is even more strikingly brought out by the final contrast between physical and biological science to be discussed here. The physical sciences have rigidly excluded teleology, the principle that the end determines the means, that the result is retroactively connected to the cause by a factor of purpose, or that usefulness is in any sense explanatory. In one of its most limited and vulgar, yet most widespread, forms this became the belief that the earth and all that is within it were intended for the benefit of man. The physical sciences had to discard such ideas completely in order to become sciences and to exclude superstition. But it is just here, as I have already mentioned in passing, that biology most clearly started on the way to a crisis.

In physical sciences it is not legitimate, indeed it is downright silly, to ask what things are for or what good they are. The question "Why do hydrogen and oxygen combine to form water?" can be answered legitimately in terms of nuclear structure, valence, bonds and other immanent properties, but not in terms of function or usefulness, not, for instance, by saying that water is formed in order to slake thirst. But in biology it is not only legitimate but also necessary to ask and answer questions teleological in aspect, concerning the function or usefulness to living organisms of everything that exists and that occurs in them. Again the insistence by many biochemists and molecular biologists that the antiteleological philosophy of physical science be applied also to biological science meant rejecting the most characteristic and important aspects of that science, widening a foolish rift, and contributing to the crisis.

In biology it still is incorrect to say that water is formed in order to slake thirst. It is, however, correct, meaningful, necessary, objective and scientific to say that animals drink water in order to slake thirst, that their digestive systems and enzymes exist and operate as they do in order to supply materials and energy, and the like. The structures and processes of organisms are useful, they perform functions, and they would not exist or occur if that were not true. They can never be understood or explained by the most complete and exact specification of the chemical reactions involved. Those reactions themselves are meaningless except as they relate to the organisms and populations and ecosystems in which they occur.

It is of interest to note, more or less in passing, that the science of psychology has sometimes suffered from a similar schism. On one hand are behaviorists, brass instrument or physiological psychologists, descriptive and objective but in a typological and idealistic way, intentionally ignoring individuality, personality, historical factors, purpose, and teleological aspects of psychology. They virtually exclude the psyche from psychology, and their understanding and explanation of what they observe, being limited to the physical level, are grossly incomplete. Yet for a time and in some quarters, this necessary but insufficient approach was considered the only adequately scientific one, just as extremists among biophysicists and biochemists have claimed a monopoly on scientific biology. In psychology as in biology, the limitations of such an approach are transcended by an organismal science (although the psychologists do not call it that), including clinical, individual and introspective methods, with attention to personality, histories, variation and teleology.

Returning to biology, recall that molecular biology is older than evolutionary biology. Nevertheless, evolutionary biology has more rapidly approached the solution of some of its most essential problems. In view of the hierarchy of organization, it would seem logical that investigation start at one end, and probably the basic physicochemical end, and thence work toward the other. As a matter of fact, that is not what has happened. The study of evolutionary biology at the organismal and populational levels did not wait on knowledge of its molecular substrate and was far advanced before the latter caught up. Darwin took it as a fact that populations include variant individuals and that some, at least, of their variation is hereditary. That is indeed a fact, and on that basis Darwin could erect a valid, although admittedly still somewhat superficial, theory of natural selection without knowing how heredity occurs or how variation arises and is perpetuated. Absence of that knowledge was, of course, a limitation, as was felt by Darwin. He proposed a hypothesis on heredity and variation that has turned out to be completely wrong. The interesting point here is that those errors did not in the least vitiate the theory of natural selection as far as Darwin had carried it. Its main features did not and still do not depend on the genetic substrate.

Nevertheless, some knowledge of that substrate was necessary in order to understand the mechanisms of natural selection at a more basic level and to refute certain apparent alternatives. That knowledge was largely provided starting in 1900 with organismal or, as it is usually but somewhat equivocally called, Mendelian genetics. As a matter of fact, most of the early organismal or Mendelian geneticists, up to about 1930, misunderstood the bearing of their discoveries on evolution and specifically on natural selection. Correct understanding came later with the extension of genetic principles to the next higher, populational level of the hierarchy. Here again the processes could be worked out in great detail at those levels without awaiting increasingly belated knowledge of lower levels. Interpretation was in terms of the gene, which was then an operational concept, the whatever-it-is that produces observed effects in organisms but whose own physicochemical nature was unknown.

The organismal observations further permitted working down to the cellular and organelle levels but not until much later to the molecular level: it was established that genes, as usually defined, reside in chromosomes and have a determinable linear sequence in them. Still without knowing what the substance of heredity actually is, many of the most important then outstanding problems of evolutionary biology could be largely solved on that basis and at those levels. For example, the processes of variation and selection were greatly clarified and extended in detail by determination that the units of heredity are discrete (nonblending), that they nevertheless form an interacting whole, that the origin of hereditary variation is in gene and chromosome mutations and recombinations, and that genetic equilibrium is maintained by random breeding and changed by selective breeding.

Finally, studies of heredity and development from cellular to individual (organismal) levels indicated that nucleic acids are implicated in transmittal of heredity or, in one variety of recent jargon, of information. With that clue, the biochemists and molecular biologists took off, and the rest is history.

The demonstration that evolutionary biology had reached many of its goals before molecular biology caught up may seem to contradict the thesis that understanding and explanation at any level of organization involves other levels both above and below. What it indicates is that there are processes characteristic of a given level and that they may, indeed to some extent *must*, initially be studied at that level. It is only as such processes become known at their own level that fuller understanding and explanation can be achieved by reference to other levels. Then the physical workings, the mechanisms involved, can be clarified by reduction to lower or more basic levels, especially from populations and individuals to molecules. At the same time, the significance of the molecular level, the origins and functions of the chemical substances and reactions involved, can finally be approached through composition to higher or more complex levels, especially those of individuals and populations. That is where we stand today, and that is how the crisis in biology can be and is being resolved.

The justly heralded recent accomplishments of molecular biology that are also of greatest interest for evolutionary biology are especially: first, determination of the structures of nucleic acids, DNA and RNA; second, methods for determination of amino acid sequences (primary structures) of peptides, polypeptides and proteins and the determination of a number of such sequences; and third, determination of the equivalents in nucleotide and amino acid sequences, or of the so-called nucleic acid codes. The interest of these discoveries is, of course, that a gene turns out to be a segment of DNA and that its primary function, or, at any rate, one of them, turns out to be the specification of the sequence of amino acids in certain large molecules. In these activities the gene, if you like, or the DNA, if you prefer, has two major functions. First, through its capacity to control its own replication it passes on to offspring cells and offspring organisms its specifications of amino acid sequences. Second, by means of those specifications and through interactions with other molecules and structures in the cell it influences not only the development but also the developed physiology of the organism.

Those discoveries are a great accomplishment, and the biologists and chemists involved can certainly be forgiven for exaggerating their importance or the distance gained toward a significant goal. That they have learned "the language of life" has a certain, quite limited metaphorical sense. That DNA is "the secret of life" is less true (as one of the stormy petrels of biology is wont to say) than that life is the secret of DNA. DNA crucially affects the cell in which it occurs, but its action is equally or more affected by that cell. Life has many secrets, and it is not yet clear how far this takes us toward the greatest of them.

Even if DNA is the language of life, we are not yet reading it. It is as if we had finally learned an alphabet, with some minor uncertainties, and perhaps could read a word or two, here an "and," there an "or." That is still a childish accomplishment and a very long way from reading a message. When finally we can read, we can proceed to the real questions: whence came this message and what does it mean?

Now the fun is over and the hard work begins. Now we must try among other things to learn what the messages are and how, so to speak, they are read by the organism. Some of the many profound mysteries remaining in genetics can be exemplified on both the molecular and the organismal sides. Is it conceivable that simply specifying a set of enzymes suffices to direct the formation of a tree or a man? It is not conceivable to me, at least. There are points, sometimes ignored, that excite curiosity. Development starts with a cell already highly organized and containing much more than DNA, or than the whole DNA-RNA-enzyme system. The DNA is necessary, but it surely is not sufficient. Even so, there seems to be too much DNA, more than should be needed to specify the primary structures of the proteins present. Is some of the DNA simply superfluous or sitting around in case of need? Evolution does not act that way at organic levels. Why should it do so among the molecules? Or consider some of the findings of organismal genetics. Genes are known operationally that modify or even eliminate the wings of flies, but no one really knows what causes flies to have wings in the first place. Or take the human disease with the interesting name of phenylketonuria, due to absence or faulty specification of an enzyme. One result is mental deficiency, usually extreme. The absence of an enzyme accounts for the deficiency, but certainly the presence of that enzyme does not account for normal intelligence. What does? The fact is that we do not know. We really do not have the least idea, because simply to say that normal information in inherited DNA results in normal intelligence is to conceal ignorance, not to dispel it.

From here on the molecular biologists have a lot of interesting but plodding work to do in accumulating more knowledge at their level, just as the evolutionary biologists, although they are further along, still have at their levels. For both, however, the exciting thing now is the connection among levels, both up and down, both compositionist and reductionist. That is what makes the plodding worthwhile and will finally add functional significance to catalogued facts.

One line of connection has already been suggested. Working upward from molecules, we need to know to what extent DNA really does specify the kind of organism that will develop and how these specifications are carried out. Very little progress has been made with those problems, but hopes can be high. Working in the other direction, we need to know where the information came from, why it is there. Obviously, this cannot be solved at the molecular level, because unless events there are countered by other factors, they tend inexorably to loss of information and the breakup of the organized system. We do already know where the information comes from: the factor that both counters the molecular tendency to lose information and produces information is natural selection. This operates primarily at the level of populations. How it operates involves also individual organisms, organs, and so on downward or inward. It involves molecules also, and among these specifically DNA, but the connection here has not yet been made in any but the most superficial way. The contributions of DNA studies to *biology* (that is, to comprehension of the life of organisms) are still more potential than actual, but surely they will come.

The fact that the contributions of DNA studies to the mainstream of biology are still largely unachieved is not always admitted by the more conservative molecular biologists. One of them recently claimed for that field the recent discovery of feedback in individual physiology and in teleological aspects of evolution. Although some of the jargon now used is fairly recent, the phenomena here in question were made known by organismal and evolutionary biologists more than a century ago: individual feedback especially by Claude Bernard and population feedback especially by Charles Darwin. It is, in fact, just the role of DNA in these phenomena that is not really clear, and it is here, in bringing DNA and other molecular studies into the whole biological picture, that the potentiality of those studies becomes important.

The increasing incorporation of molecular biology into the main body of biology proper has another currently fascinating aspect. That is the evolution or phylogeny of certain macromolecules and organic reactions such as hemoglobins, cytochromes, insulins, nitrogen excretion reactions, and an increasing number of other molecular families and cycles. Up to now the molecular biological side of such studies has mostly been descriptive: determination of differences in primary structures of molecules or in metabolic pathways in various species. Such interpretation as has been achieved is so far mainly effected by inserting these findings into phylogenies reconstructed by organismal biologists on nonmolecular evidence. The results are still quite restricted, partly because of a certain natural tendency for biochemists to carry over the typological premises of physical sciences into this inappropriate field. The data are inadequate; for instance, a single individual may be taken typologically to represent not only a whole species but even a whole order or class. The processes of evolution in populations are often misunderstood; for instance, there has frequently been an assumption that an individual molecular mutation magically forthwith characterizes a great systematic unit, completely overlooking the necessity for it to spread in populations. There are problems from the other side, too, for evolutionary biologists often do not know and occasionally do not care about the increasing interest of molecular biologists in evolution. Nevertheless this *rapprochement* is occurring, and as it does the crisis is passing.

At this happy juncture, the most important thing is to retain or, where necessary, to recover a proper balance of biological levels and specialties in teaching and research. There are fads in teaching as in so many aspects of life. So we may find elementary students in mathematics being taught set theory rather than arithmetic, which is old stuff but all the more necessary for that, after all. Or biology students may be taught recent and esoteric bits of biochemistry but not the systematics, physiology and ecology of organisms. In research, too, fads may prevail for a time, especially if staff and facilities are limited. The fad attitude may suggest that some field is now dull and should give way to another where there is an exciting breakthrough. In some quarters that was, of course, the reaction to the DNA breakthrough. That breakthrough, however, is now over, and we need not worry much about the imbalance that it caused. The next breakthrough, to call it that, the one now in progress, is between molecular and evolutionary-organismal biology, to the great benefit of both.

CHAPTER TWO

The Scope and Limits of Biology

"Biology implies biochemistry, but not the other way around." —W. A. VAN BERGEIJK

The question "Is biology an autonomous science?" was recently put to a number of physicists, biologists, and historians of science. My own immediate answer was a resounding and a decisive "Yes and no!" What is meant by "autonomous"? If we mean a definable subject including matters not ascribable to any other science, then the answer is "yes." If, on the other hand, we mean a subject complete in itself, comprising full description and explanation without recourse to any other science, then the answer is surely "no."

Neither of these answers is quite clear-cut, and either can lead to equivocation and debate. Biology is the science of life or of living organisms—a redundancy, because "organism" has come to mean "living thing" and life occurs in no other form. There is rarely any question as to what is and what is not an organism (which includes remains of formerly living things), but that characterization does become fuzzy at the edges. There is no full agreement as to whether viruses are true organisms, and the question can even arise in connection with certain macromolecules which, if hardly alive in themselves, yet never occur naturally outside of association with organisms. And then there are such subjects as sociology, which certainly deals with organisms and could with complete logic and considerable advantage be considered simply as a special case of the biological subject of animal behavior, and yet which is almost always severed from biology by its avowed practitioners and by academic protocol.

Another kind of fuzziness arises from the fact that many studies that do not deal with organisms, at least not directly, are considered biological and are carried on in departments of biology. There have even been local tendencies for them to become the main or only serious concern of such departments, tendencies rather exacerbated than balanced by other local decisions to set these studies up in separate departments of their own. I refer, of course, to the various subjects included in the subsciences of molecular biology and biochemistry. It would be absurd to say that these subjects are not relevant to biology, and it would be equally absurd to claim that they are not basically chemical. This has been one reason for denying that biology is autonomous.

If this were only a question of defining the word "autonomous," it would not be worth discussing. (Moreover, none of the definitions of the word that I find in various dictionaries is quite relevant to the point here at issue.) What underlies the ambiguous answers to the question of autonomy is the old question of reducibility. Is biology fully reducible to physical science, or to physics and chemistry? That question has involved—one could say that it has threatened—not only the autonomy but also the integrity of biology. In the threatening sense it is fast losing, if it has not quite lost, its potency. In a more modern and moderate form it is of greatly enhanced interest, and it points out a great frontier for research, and, inevitably, also for philosophy. The question now is, "To what extent is biology reducible and how are its reducible and irreducible aspects related?"

Perhaps we can best begin to delimit the field of discussion by taking two examples, equally far-out but in opposite directions. The following statement is by a professor of physics who has done considerable research in molecular biology: "In fact every one of my papers on molecular biology [several are here cited] ends up with a discussion on evolution. Yet evolution is discussed on a molecular basis without organisms (what a shock!) and is completely within the realm of physics and chemistry. It is reducible, not independent." That statement is utter nonsense, which is the reason for quoting it as exemplifying one extreme. It is absurd scientifically, because organisms cannot be, and by necessary implication are not, omitted when the molecules in question could not exist except as products and parts of organisms. It is absurd semantically and philosophically because "reducible" and "independent" are not antitheses, and neither reducibility nor lack of independence are sequiturs as the argument is presented.

The other far-out statement or overshot as we zero in on a target will be recognized as coming from an eminent botanist, E. W. Sinnott: ". . . Matter is not master. If in the universe there is an organizing principle, it may be that some of it dwells in each of us as his own soul—not as a transient and temporary configuration in atoms and molecules and quanta, but part of an eternal, universal Spirit. . . . It is a magnificent hypothesis, and, as any good hypothesis should do, it accounts for many facts that otherwise would be quite meaningless." Nothing from that source can fairly be called absurd, but this extreme is also unacceptable as science, at least, whatever one may think of it as philosophy or theology. To explain by an "organizing principle" is an egregious example of the naming fallacy, and calling this "soul" exaggerates rather than moderates the excess. The hypothesis does not in fact account for anything but only begs the question. Moreover, it is not "good," let alone "magnificent," not, at least, as science, because it has no observable deductive consequences and is open to no objective tests. All vitalistic hypotheses suffer in greater or less degree from this fatal defect.

To clarify the situation we need to avoid both the extreme of vitalism and the opposite extreme, for which I know no exact epithet and which I might call "physicism." ("Mechanism" and "materialism" are not precise or readily understood alternatives.) This can best be done not by quoting *ex cathedra* statements, such as those I have just given, but by considering a specific topic. None can serve better than the one that has now long starred in biological circles: the problem of heredity and of biochemical intracellular control, now symbolized by those magic letters DNA.

A great deal was known about heredity before the rise of modern genetics with the rebirth of Mendelism around 1900. It is not correct to conclude that nothing was known about the organismal aspects of heredity when false hypotheses were rife and before any progress had been made toward knowledge of its material basis. Mendelism in the strict or primitive sense was a start in that direction, but it still accomplished little significant reduction below the organismal level. It did lead to the concept of a theoretical-operational something or other, eventually called a gene, recognizable by its effects at the organismal level. These something-or-others took another step toward material objectivity and in reductionism when they were localized in the chromosomes, objects that are readily visible with ordinary microscopes but that are minor parts of whole organisms. The genes themselves were still operational hypotheses and were not visible, although a few thought they were; they still are not, even with electron microscopes. Yet another step soon occurred when it was found possible to locate many of the genes (far from all in any instance) in sequence within a chromosome.

At that stage in investigations it was rather widely assumed, or hypothesized, that the genes were chemical units, probably protein molecules or parts of them. Next came the discovery that the carrier of heredity in the chromosomes is not protein but deoxyribonucleic acid, or DNA, which by this very fact is demonstrated not to be a single chemical substance but an enormously varied family of polymers or macromolecules. I mention, because it cannot be repeated too often, that those substances had been known to chemists for a long time before their functions or reasons for being were discovered and that the discovery was due to studies of *organisms,* not just as chemical substances. In fact chemists do not, and I agree that strictly as chemists they *should* not, think in terms of functions for chemical substances or of reasons for their existing. That is where truly—if you like, even autonomous—*biological* thinking takes over, another point that cannot be repeated too often and that I shall indeed repeat.

Next, still traveling in the reductionist direction, one wanted to know just what composes DNA molecules, how they are put together, what shapes they assume, and how new ones arise. That, too, was discovered, not even now in fullest detail and variety but yet in most essentials.

Now came the most crucial step of any up to this point. It is usually and most easily considered in terms of an analogy, as a matter of "communication," a "message" in "code," "decoded" or "read" in certain ways. It is important to remember that this is in fact an analogy and that analogies need to be viewed with suspicion, kept firmly in their places, and thrown out as soon as they cease to be strictly parallel with the phenomena they are meant to portray or to elucidate. With that caution I shall adopt the analogy temporarily and even carry it further, likening the sequence of discoveries to the discovery and deciphering of an initially unknown language. One must first know that the language exists; heredity does occur. Next one must know where texts are to be sought; in the chromosomes. Then one must find and recognize writing; it is sequential not in proteins, as first surmised, but in DNA. To keep the next step analogous, we must assume that what we have found written is not in an alphabet or syllabary, that it does not directly represent the phonemes or syllables of the language but has been transformed into a code at least one step removed from straight language. The code was found to be in nucleotide triplets, differentiated by four bases taken three at a time, giving sixty-four different possible code signals.

Next the code signals have to be matched with the phonemes or, as turns out to be a better analogy, the syllables of the language. What the signals ("codons" is the technical jargon) match is amino acids, and most but not quite all of these matches are now well established. In passing, it should be noted that one of a number of interesting complications is that there are far fewer amino acids involved, usually only twenty, than the sixty-four possible code signals, so that some of the latter do not really code an amino acid, or two or more code the same amino acid, or, as seems to be the case, both alternatives occur. Next the syllables have to be assembled into words, and here the biochemical analogue of a word is a protein, or as much of one as is determined by one gene. The word as a whole, and not only its separate syllables, is coded in the DNA, because the code signals have a linear sequence corresponding with the sequence of amino acids in their polymer (polypeptide) form in a protein. Another complication to be noted in passing is that in the cellular "language" the average "word" is believed to have about 1,500 "syllables."

At this point the chemist or perhaps even the strictly reductionist molecular biologist may be allowed to think that his task is done. He has reduced heredity and the top-level management of cell chemistry about as far as it can go. He can sit back and write books about "The Secret of Life" or "The Language of the Gene."

For a biologist, however, this is just where things begin to get really interesting. We will have to jettison the linguistic analogy soon, but we can still use it here. The code has been located and cracked. We know how to read words. Not very many have been read, as a matter of fact, but ways to read them are known. It also happens that at this point it is generally more practical to read backward from the word to the code than forward from the code to the word, but that is beside the present point. This point is that after all that work, and the grants and the promotions and the prizes, we still are not reading the language! Even if we had transcribed all the words, which we haven't by far, and even if we knew what all the transcribed words mean, which we don't by far, we would have little to boast about as linguists if we still didn't get the message.

As a matter of fact, when we go on from amino acids and code signals for them to assembled proteins, and hence to further energy reactions and molecular syntheses mostly directed by those proteins as enzymes, then we have already turned a corner or even reversed our course 180 degrees. We are no longer reductionists tracing substance and action as far as possible to physical-chemical phenomena. We have become compositionists, putting those phenomena together in the framework of the living organism. In other words, we have ceased to be chemists and have become biologists. This, for a start, at least, indicates something as to the scope and limits of biology.

If we now take a great leap forward in the compositionist direction, we come to an extraordinary conclusion. The message that has *not* been read by transcribing code signals into syllables and syllables into words is always known in considerable detail and readily knowable in all detail but *not* in those terms! The message of heredity consists of instructions for the development of an organism and its structural and functional morphology at each stage. The message of biochemical controls and reactions adds up to the physiology of the organism. The reading of these messages by the organism is not simple, but it has been observed (with increasing skill) ever since the time of Aristotle. Recent knowledge of DNA hardly helps at all. We really do not know why or how the specifications of a particular congeries of DNA molecules and a specific set of corresponding proteins produce in one case a rosebush and in another case a man. *That* is the "secret of life" and the "language of the gene," and DNA in itself is obviously not the answer. It would not be an answer even if we knew the composition and structure of all the DNA in any organism, a goal so remote as to be not even in sight at present. No answer ever will be reached if only reductionists seek it, or if, as has been true of some workers, the DNA-RNA-enzyme system is considered the answer. Fortunately, the trend now is toward a more balanced or indeed to a predominantly compositionist attack on these problems.

Here I would like to insert a more or less parenthetical and yet important remark. Current opinion that specifications of a complete organism, its development, structure, and physiology, reside wholly or largely in nuclear DNA or in the chromosomal genes presumed to be segments of DNA is more a matter of faith than of knowledge. Control by chromosomal genes or by DNA is indeed demonstrable for a number of specific reactions and syntheses at the molecular level. In some instances also consequences of events at that level can be determined at the organismic level. But in no case does this even remotely explain how the organism came to be, or its main characteristics as an organism. Examples are provided by known genetic factors, such as that responsible for phenylketonuria, which result in mental deficiences. It is true in a certain sense, at least, that the presence of a particular allele (or other genetic factor, such as an extra chromosome) explains the observed malfunction. It does not at all follow that the absence of such a factor, or the presence of its normal alternative, explains the normal functioning of the brain. Still less does it account for the fact that the brain exists. Discovery of any number of necessary factors may not give even a hint as to the sufficient factors. As already pointed out, many genes are known that modify wing structure in flies, but nothing at all is known as to what (in a genetic sense) creates the wing in the individual. Examples could be multiplied almost endlessly.

For the most part, the important phenomena of heredity and of the manifestations of life in organisms can be studied just as well, indeed it would seem in certain instances even better, if no attention is paid to the structure and role of DNA. I think it fair to say that in this respect, as truly biological investigation and an attempt to explain vital phenomena, unmodified reductionism has failed. I am not suggesting that the failure is now complete or will be permanent. On the contrary, many researchers are now seeking to remedy it, and their currently limited successes will surely increase rapidly. The point is that this involves balancing reductionism with compositionism and a chemical with a biological point of view. Hence there is a clarification of the scope and limits of the two sciences, chemistry and biology, even where they fruitfully work on identical phenomena.

Now let us return for almost the last time to the analogy with linguistics and communications theory. We know what the message is: specification of the organism in all its aspects. We are reasonably acquainted with the code, with its transcription into a syllabary, and with the combination of the syllables into words, even though we know remarkably little about the connection of those words with the message and even though the items now recognizable as words may not convey the whole message. We still have not touched on an essential for communication, one that can be considered the most important of all from an explanatory point of view. There must be a sender, something that composes the message and some process for writing or encoding it. These analogies do indeed exist in the biological phenomena being considered, but here the analogy reaches its limit. It is not applicable to the biological process of composition and encoding of the message, these having little or no similarity to the nominal equivalents in linguistics or in straight (that is, nonanalogical) communications theory.

Everyone intelligently concerned with these matters now knows that the composition of the message is a result of evolution. For literally all of the millions of different kinds of organisms now existing the genetic message has changed through time, perhaps only moderately in some organisms among the bacteria and so-called blue-green algae, to an almost inconceivably enormous extent in others, such as ourselves. Without committing ourselves to the proposition that no other genetic changes are involved, we may agree that molecular events in DNA are an essential feature of the process. Such events can be quite varied. There can be duplications, reversals, omissions, and recombinations of nucleotide sequences, or of chromosomes and their segments, which comes to the same thing. Or there may be a substitution of one base for another in a particular nucleotide of the DNA chain. It appears that under some circumstances there can be an addition or insertion of a foreign nucleotide or segment of a polynucleotide. These molecular events are usually degenerative. (I use the word in its vernacular sense, not in reference to what is oddly called degeneracy in the DNA code.) They tend to garble messages, not to write them. In terms of communication, they are noise. If the composition of a genetic message were a molecular event and nothing else, such organisms as might ever have existed would long since have fallen apart and instead of having millions of species, so awesome and wonderful, today we would have none at all.

Most obvious of the phenomena that make total reduction of organism and evolution to the molecular level absolutely impossible is the phenomenon of adaptation. Without worrying over the question of whether all characteristics of organisms are adaptive, it is an obvious fact that most of them are. Although evolution has many other aspects, evolutionists have always recognized that explanation of the origin and maintenance of adaptation is the crucial test for any general theory of evolution. The origin of adaptations so diverse and for the most part so unlike those of primeval organisms demands that innumerable molecular events must have occurred and, far more essential, that from these there must somehow have arisen an adaptive combination, a message that makes sense. Molecular events in themselves have no such result, but rather the opposite. Therefore, there must be a process at some other level that acts on the variation generated by those events, selecting some and rejecting more, combining those selected in an adaptive way, and conserving the results.

That process is, of course, natural selection in its modern sense, which is broader than Darwin's but not contradictory. The main point here is that natural selection is not a molecular event. It is at quite the opposite end of the scale of composition and reduction. Its reduction to terms of molecules is absolutely impossible, although it literally creates molecules and that fact is necessary for understanding either natural selection or the molecules.

Adaptation is a relationship between organisms and environment that is favorable to the organism. (Anticipating slightly, we should mention immediately that it is favorable to a population or species of organisms, and only incidentally to an individual organism.) It necessarily involves hierarchic levels above the molecular and systems above the cellular. Lamarck and the Neo-Lamarckians (even though their theory was in fact non-Lamarckian) thought that the relevant system was the individual organism. Darwin agreed as to that relevance, but he observed that in respect to adaptation much the most important system is the reproducing population. It was probably the most original and essential of Darwin's discoveries, although not always recognized as such. We now know that the populational system is not only the most important; it is literally all-important.

In fact, the system involved in natural selection and adaptation is an open one including both a reproducing population and its whole environment. However, it is unfortunately not true that the wind is tempered to the shorn lamb. It is the organic population, not its environment, that becomes adapted, maintains adaptation, changes adaptation or—and historically this has been the usual case—finally fails to do so and becomes extinct. To be sure, I am leaving out some complications here and I mention a couple of them to indicate that they are not forgotten but just are not considered entirely necessary to the present discussion. Feedback between organisms and environments does occur at the individual as well as at the populational level and this probably can, although it usually does not, affect genetics, but it is not a necessary or general factor in adaptation. Further, selective reactions may occur within an organismal system without direct connection or feedback with the environment and these may affect the genetic message. They do so significantly, however, only through involvement in the populational system, where they are almost sure also to become involved eventually in environmental feedback.

I mentioned earlier that organic molecules, or one may better say biomolecules, have functions and reasons for existence, although these terms and concepts are aliens inadmissible to the realm of chemistry. Their functions have to do with the continuance of populations of organisms and with adaptations promoting that continuance. Their reasons for being have to do with the long action of natural selection within the continuing populations. DNA or any other biomolecule cannot be understood in terms of its chemical and physical characteristics alone, but only if its biological origins and operations are also considered. A slight beginning is made when it is established that DNA specifies proteins or that ATP is an energy source in exergonic reactions. We do not really understand these molecules and we are not really dealing with them as biologists until we note that their very existence is the result of organic evolution and that their functions must be followed upward through the levels of cells, tissues, organs, individuals, and populations to the all-embracing system of life and its environments before they are really understood.

Now I believe that I have expounded my title, "The Scope and Limits of Biology." The subject is fairly well boxed in. Where it meets with chemistry, it is distinguished by dealing with molecules that exist by reason of life and those viewed as parts of living things. Perhaps something more should have been said about scope and limits in the area contiguous to the social sciences. For the present, however, I am content to define the social sciences as those branches of biology dealing with organisms that have language. (I do not mean the "language" of the bees, which is not language.) A distinction between biological and social sciences is at least as pragmatical as it is logical.

I think, too, that the question, "Is biology an autonomous science?" has been answered as well as need be, even in its overtones and undertones. Within its limits, the scope of biology is broad and continuous. It cannot be divided logically, usefully or sensibly into compartments such as molecular and organismal biology. That it has in fact been so compartmented on occasion simply demonstrates that some scientists and some schools are not completely logical, useful, or sensible. Biology is integral both objectively, in its materials, and subjectively, in its aims and principles. Even when the objects and methods of study are the same, it is distinct from the physical sciences in point of view. Indeed, a legitimate distinction from the social sciences could also be made on that basis.

Biology is not fully reducible to terms of the strictly physical sciences. On this another comment is necessary. The statement of nonreducibility is too often taken to mean that the phenomena of life are supposed to contravene those of the physical world, that biology wrongly claims to need "new, independent first principles" (I am quoting from the physicist who talks about biomolecular evolution without organisms), or that there is some (perhaps divine) "organizing principle," "entelechy," or other nonmaterial something or other necessary to understand biology. Living organisms are made of exactly the same materials and are subject to exactly the same forces, laws, and principles as the nonorganic world. It can even be said that life results from the nature of the nonorganic world, certainly with no independent first principles added. But despite their diversity all living things share a kind of organization of nonorganic materials according to nonorganic, or, rather, to universal principles. Organic systems have characteristics not present in any nonliving system. One way to analyze the living system is to take it apart, but then its characteristics *as a system* are lost, and those are the irreducible features of life.

That feature of irreducibility occurs not only between biology and the physical sciences but also within the biological hierarchy.

The phenomena of communities cannot be explained *without* and yet cannot be entirely explained *by* those of its separate species. The phenomena of species cannot be explained without but cannot be entirely explained by those of individuals. And so on down to the phenomena of cells (or of organelles), which cannot be explained entirely by those of molecules. Finally, such statements are just as true if made in the opposite, the compositionist direction. The phenomena of molecules cannot be explained entirely by those of organelles or cells. . . . The phenomena of species cannot be explained without but cannot be explained entirely by those of communities.

Biology is both reductionist and compositionist, both nonreductionist and noncompositionist. Its adequate study requires equally reduction and composition. Let us quit an argument that turns out to be rather silly and accept the reality of life as integral in all its aspects.

CHAPTER THREE

Organisms and Molecules in Evolution

"Life is the necessary and sufficient condition for macromolecular systems, but macromolecules, though necessary, are not sufficient for life."

—W. A. van Bergeijk

It is universally recognized that molecules of biological importance may evolve—that is, they may change in the course of time as have the organisms in which they occur. Some molecules, like adenosine triphosphate, are so nearly universal and invariable as to suggest no evolutionary sequence, but many others surely have evolved, notably groups of proteins and, obviously, DNA. Before the importance of DNA was known, Florkin had already discussed the systematics and evolution of various families of molecules. In such instances evolutionary interpretation of the biochemists' findings requires information from paleontologists and systematists, information especially on the time scale involved and the phylogeny and relationships of the species in which varying molecules are to be compared. An example is the hypothesis that serum proteins or cytochromes have changed in a regular if not linear manner with respect to time—that they have evolved by some sort of internal constant-rate mutational process and not in an irregular or a specifically adaptive way. In fact, when the data are replotted with what seem to be the most probable time co-ordinates they indicate that the hypothesis is incorrect or, at least, that these data do not support it. The hypothesis has since been modified, but it exemplifies the clarifying confrontation of molecular and organismal data.

Other interesting examples of such confrontation arise from further studies of serum proteins, such as some by Goodman. Phylogenetic relationships of the animals concerned, primates in Goodman's case, are inferred from the apparent degrees of homology in their various serum proteins. The lineages thus inferred then permit conclusions as to the evolution of the proteins themselves. Similar inferential methods have been applied to the evolution of hemoglobins, also in primates, by Hill, the Buettner-Janusches and others. When phylogeny is inferred from the molecular data and molecular evolution is inferred in turn from that phylogeny, there is an element of circularity, which does not wholly invalidate the method but does warrant some reservations. A necessary cross-check is to arrange the molecular data in the framework of a phylogeny based entirely on nonmolecular evidence. It should be mentioned in passing that this, too, has sometimes led to semicircular reasoning when molecule-based phylogeny has been compared with phylogeny with other bases: agreement between the two has been taken as the requisite validation of the molecular approach to phylogeny, but nonagreement has been taken as evidence of the greater reliability of the molecular method.

However, the most important reason for relating organismal and molecular evolution to each other is not simply the testing of hypotheses or the validation of methods. It is the balancing of points of view and the achievement of more complete explanations. Wald has said that "living organisms are the greatly magnified expressions of the molecules that compose them." Anfinsen believes that "we may almost define the life sciences as those concerned with the elucidation of the mechanisms by which molecules exert their specific actions in living cells." In fact there are many respectable and even eminent students of the life sciences who have no concern whatever with molecules or their actions. Concentration on one level of organization to the practical exclusion of others is often a necessity of specialized research, but nowadays almost everyone agrees that eventual understanding of relationships between levels is also necessary. Sonneborn has emphasized the fact that molecular genetics could only have arisen through, and would now have little meaning apart from, "classical" or Mendelian organismal genetics. Weiss has pointed out that there is a "cellular control of molecular activities" as well as a molecular control of cellular activities. There is also an organismal control of cellular activities, and, for that matter, a populational control of organismal activities. Indeed both Wald and Anfinsen, in the works from which one-sided aphorisms have been quoted, were concerned with relationships of molecules to higher organizational levels in evolution.

The sort of problem that can arise from a limited approach is exemplified in a recent article by Mora. He points out that living organisms have a teleological or purposive aspect which he proposes to label "urge." He finds that this aspect is inexplicable at the molecular level as hitherto studied. He proposes, but does not describe, a new approach, to be frankly permeated by teleology. Although he seems to think or hope that this may still be naturalistic, he does not clearly state what a naturalistic teleology might be. Now, this is precisely the problem with which organismal biologists have been coping for generations. Unknown, it would seem, to some biochemists, they have achieved a naturalistic (or, in a sense, materialistic) explanation of what is now often called the teleonomic aspect of organisms. The teleonomic, or *apparently* teleological or purposive, characteristics of organisms are adaptations. They include "urge" itself in Mora's sense, its manifestations, and its results in the activities of individuals and the evolution of populations. Teleonomic adaptations arise in the course of evolution, and the factor governing their origin and maintenance is natural selection. That is surely as true at the molecular level as at any other. However, the ramifications of natural selection at various levels are far from simple.

The process of natural selection, as now understood, is complex rather in its concrete working and its interactions than in its basis. That basis is simply differential reproduction correlated with genotypic constitution. If some individuals in a population have more surviving and breeding offspring than others, and if there is a consistent average difference, however small, in the genotypes of those who have more and those who have fewer, that is natural selection at work. The actual selection—that is, the determination of which individuals have more or fewer offspring that survive to breed in their turn—is an interaction between environment, in the broadest sense, and the population, in all its individuals throughout their complete ontogenies. Aspects of this process are discussed at length in recent works which supply many details not given here and some of which are cited at the back of this book.

Natural selection requires, first, reproduction and, second, hereditary variation of such a kind as to influence the success of reproduction under existing circumstances. When those factors are present, natural selection *necessarily* occurs. In precellular evolution it necessarily began when there were replicating molecules that differed in the rate or efficiency of replication. However, the pertinent unit is not the replicating molecule but the reproducing system. This was presumably a molecule at first but became a cell at the protistan level, and is a dynamic unicellular-to-multicellular ontogenetic individual at metaphytic and metazoan levels. Selection acts on the whole phenotype and can single out genes only to the extent that they have phenotypic effects separable both phenotypically and genetically from those of other genes. Although selection apparently does act in an analytically separable way on some particular molecules, it evidently does not do so as a rule. It usually acts on supramolecular phenotypic characters, on whole complexes of them, or indeed on all of them at once. Since most genes are pleiotropic and most characters are polygenic, it follows that selection usually is not concentrated on single genes, as might appear from the necessarily oversimplified models first formulated by population geneticists. Although the connection is not yet well understood, this presumably means also that it is unusual (it may even be impossible) for intermediary molecules such as enzymes and other proteins to be selected for or selected against independently of other molecules.

In considering the effect of selection on particular features of an organism, it is important to judge how far these are in one direction from the genes and in the other direction from the phenotypic characters directly subject to selection. Behavior is subject to particularly strong selection, and it is probably farthest removed from the genes and also most elaborately polygenic as a rule. Some single-gene determinants of behavior are known, but they are exceptional. Proteins or, at least, intracellular enzymes are believed to be almost directly and uniquely determined by one or a few particular genes. The effect of selection will surely be influenced by the length of the functional chain from the genes to the character selected for or against. As a rule, with exceptions, the effect becomes more, not less, diffuse and less, not more, direct as the level of the gene is approached.

Zuckerkandl has argued that a molecule like hemoglobin is preferable to most "structural," or more remotely phenotypic, characters for the determination of affinities because it is so near the genes, so nearly a direct reflection of part of the DNA code. It may be added that hemoglobin is so literally vital that natural selection may here act at a level near the gene. Those are advantages in certain respects, but they are accompanied by disadvantages, and the more distantly phenotypic approach also has advantages. Zuckerkandl has shown that, "from the point of view of hemoglobin structure, it appears that a gorilla is just an abnormal human, or man an abnormal gorilla, and the two species form actually one continuous population." From any point of view other than that properly specified, that is, of course, nonsense. What the comparison seems really to indicate is that in this case, at least, hemoglobin is a bad choice and has nothing to tell us about affinities, or indeed tells us a lie. (It does show that men and gorillas are rather closely related, but that has long and more accurately been known from traditional morphological comparisons.) Of course, as Zuckerkandl points out, we should use not just one kind of molecule but many, preferably proteins. However, if one can be misleading, so can many! (Let me add that Zuckerkandl's discussion of the phylogenetic interpretation of molecular data is invaluable.)

In some respects it is a drawback that hemoglobin, various enzymes, and some other proteins are so near to the genes in the functional chain. It means that each character is genetically determined by, and therefore provides a sample of, only an extremely minute part of the whole genetic system—apparently only two genes in the case of hemoglobin and probably only one for many enzymes. The farther a character is from the genes, the more likely it is to sample a number of genes or a really significant part of the whole genetic system. The complexity of the genetic determination of a characteristic is a positive advantage, not a disadvantage, when the purpose is to determine affinities of whole organisms. Moreover, such characters are in almost all cases those which were in fact subject to selection. On an average, the farther we are from genes the nearer we are to the action of selection, and thus the better able we are to interpret the adaptive processes involved.

When, as is usual, selection is on the phenotype and well removed from the genotype, all that matters is that the genotype should in fact result in the selectively favored phenotype under the existing conditions of development. In this sense, or beyond that point, it really can be said that the genotype does not matter in adaptive evolution. There is ample evidence that genotype-phenotype determination is not unique in either direction. Phenotypes that are apparently identical and that seem to be equal in the face of selection can have markedly different genotypes. There are also many systems—genetic, ontogenetic, and selectional—that tend to channel phenotypic development in the face of considerable change or variation in genes and hence, presumably, also in many families of macromolecules. I am arguing not that any one kind of evidence on evolution—genetic, molecular, phenotypic, or other—is superior but, on the contrary, that no one kind suffices in itself.

The evolutionary study of molecules has raised a number of special problems, not always seen in the same way by molecular and organismal biologists. The phenomenon that has caused most trouble in attempts to determine evolutionary affinities is convergence: the development of similar characteristics by organisms of different ancestry. Any addition of evidence would be most welcome, especially if it involved characters unlikely to converge. Here the molecular biologists do not agree; Wald, for example, says that convergence is much more likely at the molecular level, while Zuckerkandl independently maintains that it is less likely. To me, as an organismal biologist, it seems that Wald is probably right. Convergence to the point of identity or of seriously confusing similarity would appear to be more likely in a single kind of molecule, even one as complicated as a protein, than in such phenotypic characters as are end results of the interactions of a very large number of such molecules. Sanger (with others) has shown that the insulin composition of sperm whales is identical with that of pigs and quite different from that of sei whales! To be sure, a sequence of only three amino acids is involved, and both differences and resemblances could be incidental without even true convergence, but the lesson is there. Fortunately, the fact that protein and morphological convergence may be independent of each other gives a double check if the evidence of both is available.

Another problem, discussed at some length by Anfinsen, arises from the evidence that proteins have parts that can vary greatly or even be removed altogether without seeming to affect function. There is also the concept of "dormant genes," discussed by Zuckerkandl and others. This concept is, again, related to the hypothesis of regular, secular change in molecules, mentioned in the opening paragraph of this chapter. Essentially the same question has long been discussed by evolutionary biologists, in this form: can a gene (or allele) be neutral with respect to selection? It is impossible to establish complete absence of exceptions, but so far every supposedly neutral gene that has been adequately investigated has turned out not to be neutral. There is a strong consensus that completely neutral genes or alleles must be very rare if they exist at all. To an evolutionary biologist it therefore seems highly improbable that proteins, supposedly fully determined by genes, should have nonfunctional parts, that dormant genes should exist over periods of generations, or that molecules should change in a regular but nonadaptive way.

This unsettled question could have far-reaching significance, for instance through the hypothesis, suggested but not fully supported by Anfinsen, that the invariable or fully homologous parts of proteins in different animals are the functional, or at least the most significantly functional, parts. It would then seem to follow that the actual specific differences in proteins may be little or not at all adaptive, and this again seems unlikely to an organismal biologist. However, Anfinsen also points out (and the examples could be largely multiplied from other sources) that, for instance, serum proteins with no immunochemical similarity at all may be fully and identically functional. It is certainly not true as a generalization that molecular differences among species are commonly nonfunctional or nonadaptive, and indeed I think no molecular biologist would go to that extreme.

It is undoubtedly on questions related to adaptation that an evolutionary synthesis of molecular and organismal viewpoints and data will be most useful. I shall here give briefly two further examples from work by Wald, not because I happen to disagree with his interpretations but because his brilliant studies provide such ideal data on the molecular basis of organismal adaptation. He shows that freshwater vertebrates generally have retinal pigments containing vitamin A_2, while marine and land vertebrates generally have A_1. He interprets this as a phylogenetic phenomenon, with A_2 in ancestral (true) fishes, supposedly freshwater forms, and A_1 developed in progressive phylogeny by marine and land descendants. He finds it inexplicable and almost an unnecessary complication that, for instance, reptiles, primitively having A_1, "revert" to A_2 when they adapt to fresh water. To an organismal biologist, the picture, including the apparent anomalies and supposed reversions, suggests interpretation in terms of adaptation, primarily, and phylogeny only secondarily. Many, but perhaps not quite all, of the observations would be explained if we assumed that A_2 is adaptive in freshwater forms and A_1, in land and saltwater forms—so much so that selection usually produced these adaptations rapidly and tended to erase purely phylogenetic effects. I have no idea what the difference in adaptation might be, but suggest that study from this point of view might clarify the molecular function involved.

A second example from Wald is his demonstration that tadpoles resemble fishes in a number of biochemical characteristics, whereas adult frogs have a biochemistry more like other land vertebrates. Amphibians were, of course, derived from fishes, and Wald interprets these changes as "the most striking instances we know of recapitulation." In my opinion there is no reason to invoke recapitulation and definite reason not to. As regards the species in question, it would appear that tadpoles are adapted to live in the water and adult frogs to live on land. In spite of some complications, this is the plausible explanation for nitrogen excretion: ammonia in water, urea out of it. Other changes may be less clearly adaptive but are likely, at least, to be adaptive. Some of the evidence, also given in part by Wald, is that when amphibians go from land to water, as some do, the changes tend to go in the opposite direction; they antirecapitulate!

Finally, let us turn (or return) to the structure of the whole adaptive system, its causations, and the place of molecules in it. The most basic of all molecules, in this context at least, is DNA. Its influence is exerted, in part if not altogether, through RNA. Recognizing the RNA as an agent of DNA in this sequence, we conclude that RNA is not the cause of the eventual action: synthesis of a protein. (One could raise some delicate semantic problems here, but I think the statement can stand as written for present purposes.) Then is the DNA the causative agent in a really explanatory sense? It carries, as we say, a message (another semantic problem!) and is indeed a messenger and an agent just as much as messenger RNA is. In following the chain back we reach a really significant point of causation not when we locate the message, which is in the DNA, but when we learn where the message came from to begin with, what composed it. Any message composed, so to speak, by the DNA itself would be in the language of mutation. But mutations are predominantly inadaptive, and the message, beyond doubt, is almost entirely adaptive. Mutations form what may be called letters or words, to continue the now somewhat shopworn metaphor, and in that way they supply materials that permit something new to be said and that limit what can be said. However, they certainly do not compose the message in any meaningful sense.

The message, or at very least the greater part of it, relates to interaction of organism and environment. The interaction involves the whole organism, and hence arises and expands from the molecular level. There must be some sort of feedback from the organism-environment interaction into DNA, and hence into the other molecules. There are, as is well known, innumerable feedback mechanisms at the molecular level itself, and many or most of these are responsive to interactions with the environment. The Neo-Lamarckians, before much was known about feedback or anything at all was known about molecular genetics, supposed that evolutionary feedback was of the same kind, within individuals and into the genetic system, whatever that might prove to be. Now, however, we do know about DNA and other essentials of the genetic system, and we know beyond serious doubt, even though it seems rather odd, that DNA is not subject to feedback within individuals. That is, as Pontecorvo has put it, "the *structure* of the genetic material is not subject to regulatory change . . . although the *expression* of the genetic material . . . is subject to regulation—qualitative and quantitative—at all levels of organization. . . ."

Changes in individual expression—to put it figuratively, the way the message is read—do not affect the message itself. The necessary message-constructing feedback is not here but in a system of higher order: in the population and not the individual. It operates through natural selection, which operates in populations, just as populations are what really evolve. Thus, through a different approach we come again to natural selection and now see it as the most truly causative (although not the only) element in the adaptive system. Viewed in this way, it is the composer of the genetic message, and DNA, RNA, enzymes, and the other molecules in the system are successively its messengers.

CHAPTER FOUR

Biology and the Public Good

"Increasing concern has been manifested in several quarters with the bearing of rapid advances in scientific and technical knowledge in biology, chemistry, and medicine."—Proposal of the National Research Council for a study of biological research, social behavior, and social policy

My title, "Biology and the Public Good," has been taken with special consideration of Benjamin Rush. His whole life was devoted to the application of biological principles to the public good, often with exceptional courage. His was a technological aspect of biology, that is, medicine and what we now call psychiatry. This may be balanced not against but with biology as a science, apart from its applied aspects, in the person of another historic figure connected with Dickinson College: Spencer Fullerton Baird, who was graduated there in 1840 at the ripe age of seventeen. He was on the Dickinson faculty in his twenties from 1845 to 1850 and was later the Secretary of the Smithsonian Institution. He was a pioneer student of the American fauna, and was instrumental in the promotion of biological research in general, notably through the establishment of the Marine Biological Station at Woods Hole, Massachusetts.

In the hands of Benjamin Rush's followers, biological research increasingly contributed to the public good in the applied form of medicine in the broadest sense, and, it may be added, also through agriculture, likewise in the broadest sense. This side of biology is to medicine and agriculture about as physics and chemistry are to engineering and to various kinds of manufacturing. There is much to say about that; indeed the subject can never be exhausted. Its details are esoteric, but the existence and general nature of these contributions are common knowledge, widely publicized in print at all levels of literacy, and heavily supported financially by direct donation, in industrial laboratories, and through the devious channels of government.

While duly honoring and enjoying the benefits of the achievements of Rush's successors, I am, myself, a follower of Baird more than of Rush, and I pursue biology as a so-called "pure" science rather than as a basis for medicine and agriculture. I believe that the existence and nature of contributions to the public good by what is sometimes called the impractical side of science are less generally understood. I believe that that is especially true of biology. It will further be sadly pertinent to this theme that some practical applications of biology have been misapplications and that many potentially good applications have been neglected. Even within the fields of medicine and agriculture, the interaction of scientific biology and biological technology is not altogether for the public good. In some respects the balance seems even to be for the public ill. An essential part of my thesis is, however, that the fault is not on the side of biology, the science, and that such remedies as are in our power must include more knowledge of biology, better understanding of it by the public, and a better balance of its different aspects.

After that preamble, perhaps I should at once proceed to specification of particular applications and misapplications of biology to the public good and ill. However, I have decided to consider first, at limited length, a contribution of biology that is less tangible but even more pervasive and ultimately even more important; its philosophical import, its bearing on our world view and on our human nature and concepts of common sense.

The rise of science seems, oddly and contrary to some widespread views, to have taken place almost independently of the much longer, more continuous progress of technology. The present relationship of the two, so intimate that they are hard to define separately, is a comparatively modern phenomenon. The reason for the origin of science was not in its potential applicability, but in its capacity to rationalize the world and reconcile reality with the caprices and inadequacies of the human mind. The rise of science depended first of all on belief that the universe makes sense, that its phenomena are orderly and can be explained without recourse to the miraculous, the mystical, and the ineffable. Without reference to its origin or to what might in some sense exist apart from it, the material universe came to be construed as a self-contained, self-regulating physical system.

Up to about a century ago it was almost always considered that biological phenomena were an exception, involving something quite outside the material system. Of course it was observed that living bodies are material and subject to physical forces and processes, but the life in them was construed as something quite different. A living man and a dead one need not differ in physical structure, so, it was thought, they must differ in some nonphysical life essence. The fallacy was not at first observed even when motors were invented, identical in physical structure when running and when stopped. As to the origin of living things, two views were long held: that each kind was created in the beginning and has since been self-procreated, and that individuals of many although not all kinds are continually generated spontaneously from inert matter. It was finally realized, as late as the nineteenth century, that the two views are contradictory, although even now it is not universally realized that neither one is true.

It is an old wives' tale that human nature and common sense do not change. Now it is obvious that human nature has changed, at least to the extent that it is not ape nature. If we find still too much of the ape in it, we can be hopeful that what has changed somewhat can change more. So with common sense. It was common sense that the earth is flat and is daily crossed by the motile sun. It was common sense that flies are engendered from putrid meat. These things could be seen and so could not be doubted. Although the unique event could not be witnessed, it was equally common sense and equally indubitable that mankind was created all at once, just as it is. But we know that those common-sense notions, even those based on universal eyewitness, do not correspond with the objective world. They thus cease to be common sense, and it becomes common sense that the earth rotates daily and revolves yearly, that flies arise only from eggs laid by flies, and that man has slowly evolved from ape, reptile, fish, invertebrate, protist, and ultimately from simple matter. Science is our principal means of thus changing common sense and bringing it into closer (hence, too, more workable) correspondence with the objective world. The greatest contribution of biology to the public good is the discovery that living things also belong wholly within the self-contained, self-regulating system first delineated by the physical scientists. In fact, most recent advances of applied biology in medicine, agriculture, and elsewhere depend on that most fundamental of biological principles.

It is curious and indeed distressing that this development, which eliminated paradoxes and unified our concepts of the universe, nevertheless has produced new paradoxes and new conflicts of concept. In some quarters, instead of expanding the scientific world view of conceptual framework to include biology, there has been a tendency to restrict biology to the framework already established for the physical sciences. Because biology was obliged to reject vitalism and to become material and objective, it was felt by some that its principles and explanations must become strictly and solely physical and chemical. In philosophy this view was for a time almost universal and is still usual: "philosophy of science" meant "philosophy of the physical sciences." If biology was brought in at all, that was only to apply to it a philosophy neither derived from it nor appropriate to it. In biology itself a result was the tendency to split into two camps, the molecular biologists with a reductionist approach, considering biological principles and explanations in physicochemical terms only, and the organismal biologists, dealing with living systems as such. Only recently are there healthy signs that the monomania of philosophy of science and the schizophrenia of biology may be curable.

Of course it is nonsense to consider living systems as no different from or even as wholly reducible to nonliving systems, just as it is nonsense to think that living systems must include something non-objective and exempt from material law. I consider these matters so important for the public good that I want to spend just a few minutes contrasting the limited outlook of the physical sciences with the broader but still objective outlook necessary to include also the life sciences.

In the physical sciences the systems involved are relatively simple, even in principle, simpler still in the form in which they are usually studied. Living systems, that is, organisms, are incomparably more complex. To study these systems it is possible and necessary to analyze them into simpler components, for example to study intracellular chemistry one reaction at a time. However, the system as such, that is, the life of the organism, cannot be understood in terms of any or all such reactions taken one at a time but only by considering all at once within the structural and chemical system as a whole. An analogy might be derived from a TV set. Understanding it requires knowledge of the characteristics of each separate component, but understanding is incomplete to the point of inanity unless one also considers how the components are put together and how they interact. (Remember, too, that the simplest living system is vastly more complex than the most intricate TV set.)

That analogy breaks down in relation to another contrast of physical and biological science. In biology it is fruitful to study a system as a whole, even if knowledge of its components is limited or nil. For example, Darwin correctly inferred the essential action of natural selection on heredity before anything effective was known about genetics, and geneticists worked out essential characteristics of genes before anything effective was known about DNA, which is the chemical basis of the operational genes. In fact, that worked the other way around: it was the organismal study of Mendelian genetics that led to the discovery that genes are segments of DNA, and not the molecular study of DNA that led to the discovery of its genetic function.

In physical science the postulates and methods are typological. It is usually assumed that objects studied, particles, molecules, etc., fall into a limited number of types or sets and that no notice need be taken of differences among individual objects or events. Even statistical study may not be considered as involving individual differences in histories or characteristics but rather as a reflection of a statistical uncertainty inherent in the universe. But organisms cannot be truthfully or usefully considered in terms of types. Variation is essential to their nature and is built into their populations and systems of heredity. The community processes of living nature and especially the progress of evolution are possible only because populations of organisms are not reducible to types. Statistical methods, if used, are means of describing, summarizing, and inferring the variations of populations, not of ignoring these or of coping with inherent uncertainty.

With rare and marginal exceptions, the physical sciences are nonhistorical. It is generally assumed that the phenomena under study are indefinitely repeatable and do not include historical changes or unique events. On a larger scale gravity or on a smaller scale intranuclear forces are assumed to have no history, to be the same regardless of date or duration. It is further assumed that phenomena are predictable, that fulfillment of prediction is the essential test of hypotheses, and that a confirmed prediction is the same thing as—and also the only valid form of—explanation. That is, if certain conditions lead to a subsequent, predictable result, one can change the tense or the direction of inference and say that the observed result is explained by the preceding conditions.

With, again, marginal exceptions, those assumptions are necessary to physical science; it can hardly proceed without them. They have hence been considered basic to the philosophy of science in general, and yet none of them applies strictly to biology. Organisms have histories. What they are and how they work do not contravene or transcend the timeless principles of physical science, but they also depend on the whole sequence of past states in the individuals and in their ancestors for (at least) several billions of years. Moreover, each of those states was unique, distinct, and unrepeatable in essence. Evolutionary changes are imperfectly or not predictable, certainly not predictable in clear detail in practice and probably not even in principle. A current state can be explained by looking backward to earlier historical states, but such explanation is distinct from and rarely accompanied by prediction.

Finally and perhaps most importantly, physical science not only eschews but also prohibits concepts of teleology or purpose. It is antiscientific to ask why an unsupported object falls, if "why?" is used in the sense of "What's the good of it?" or "What's the purpose?" To be sure, such questions sometimes do have answers. To the question "Why is the Grand Canyon?" the answer used to be "To throw used razor blades into," and is now "To build dams in for the benefit of Arizona politicians and realtors." Here is just the point: these are really *teleological* answers. They make sense (even if facetious sense) because the teleology, the purpose, is in terms of living organisms: men who use blade razors, politicians, and salesmen.

In fact, in biology it is not only legitimate but also necessary to ask teleological questions if life and its manifestations are to be understood. This is not a return to the Aristotelian nonsense of the end setting the means or the effect determining the cause. It is the simple and requisite recognition that biological phenomena do have functions and are determinately linked to their functions. The ADP-ATP cycle of the biochemists does produce energy that is useful to organisms, that serves a purpose in the economy of whole living systems. Intracellular—or for that matter extracellular—molecular phenomena remain completely meaningless from a biological point of view until their functional significance is discovered. DNA was long known to the chemists and evoked no special interest until organismal biologists discovered its biological function, its significance for whole organisms and for populations of organisms. It was Darwin's demonstration that the functional, apparently teleological aspects of living things can be explained by their material history that finally brought biological phenomena into the self-contained, self-regulating system of the material universe. A philosophy of science is not only inadequate but also false if it ignores or rejects the fact that teleology, in the Darwinian redefinition, thus became a part of the scientific world picture.

In turning from those broadly theoretical and philosophical matters to more concrete aspects of the subject, it may seem that there is in fact a radical change of subject. There is, nevertheless, a direct continuity in at least two ways. In the first place, accurate concepts in science and their successful application depend largely on their background, both theoretical and philosophical. The recent reaction against Lysenkoism (or as they say, Michurinism) in the Soviet Union, although technological on one level and political on another, is more significantly a cryptic philosophical shift. In all countries the progress of biology, both pure and applied, necessitates the broadening of the philosophy of science along the lines that I have sketched.

The second conceptual connection between my topics is in the fact that the unity of a comprehensive philosophy of science depends on the hierarchy of organization. This hierarchy runs more or less from the subatomic through atoms, molecules, cells, tissues, organs, individuals, specific populations, communities, and biomes, comprehensive biological systems, to the whole realm of the organic and its environments in space and time. The phenomena of life in full array occur only from the cellular level upward, and the science of biology therefore starts there. Lower levels, from molecules down, are the physical substrate of biology, absolutely necessary for that science but not, strictly speaking, a part of it. We have seen that the inadequacies or distortions of much that has been advanced as philosophy of science stemmed in the main from failure to take unbiased account of the higher levels of organization, those peculiarly biological. We shall now see that failures and misapplications of biological technology have often had a similar cause: failure to take account of problems involving or arising in neglected higher levels of the hierarchy of organization.

Since the field of medicine has witnessed the greatest triumphs of biological technology, perhaps the most significant example to start with here is one in which the triumph begins to have a hollow sound. The various so-called wonder drugs, those usually tolerated by humans but usually fatal to various disease-causing microorganisms, have indeed worked wonders in combating such diseases. However, it happens more and more frequently that continued use of such drugs, instead of overcoming the last remnants of infection, causes the evolution, through natural selection, of strains of pathogens (disease-producers) highly resistant or immune to the drug in question. It has even been found recently that in some instances resistant strains thus evolved can pass on resistance to microorganisms that have not been exposed to the selective action of the drug. Some despairing medical researchers have expressed fear that our whole armament of germicides will become useless because they are unable to cope with agents of massive infection and contagion whose evolution they have themselves caused. This branch of medicine would then be set right back to its beginning and would have to start again on some other basis.

The problem here seemed quite simple when it was viewed at its lowest hierarchical level and in typological terms. The germicidal molecule interfered with (in all likelihood) a single vital reaction in the individual germ and so ended the disease. But microorganisms are not types. They occur in immensely numerous and highly variable populations. The evolution that occurs at that level—undesirable from the medical point of view—is inevitable and could, indeed, be foreseen by attention to the pertinent higher hierarchical level, that of populations rather than individuals. One may now predict that the way out by a new approach will take account not only of that level, because events have already made that unavoidable, but also of the vast and intricate interactions at levels from atoms to communities with special attention (as an absolute minimum) to populations of humans, of pathogens, and of carriers of infection. Thus, only more biological knowledge on broader fronts can lead us out of the impasse into which a little narrow biological knowledge led us by causing us to produce essentially new diseases with which we cannot cope.

The many readers of Rachel Carson's *Silent Spring* are familiar with another set of intensely serious problems in applied biology. The widespread use of chemicals to control noxious insects has had a whole chain, or many chains, of side effects on other animals and on man, ranging from merely undesirable to downright disastrous. Thus the attempt to solve one problem of applied biology has caused numerous new problems. Again, the initial problem was attacked on too limited a basis and not on all appropriate levels. This is no simple matter of a harmful insect and a poison to which it is susceptible. (And incidentally here, analogously to the story of germicides, the result sometimes has been to evolve immune insects.) Bringing poison into the environment and attacking certain members of the community does not affect just those members. It changes the whole ecological setup and has repercussions, usually undesirable, through the whole community and out beyond a particular community. The problem should have been studied, and now it *must* be studied, in the broadest way and up to the very highest level of the hierarchy of biological organization.

Carson made a strong case, I think an unanswerable one, that in many instances the balance of the effect of use of insecticides has been for the public ill. Yet the argument is vulnerable to the criticism that it, too, fails in judgment of the total ecological situation. It remains true that importantly noxious insects must be controlled and that any form of control will necessarily bring about widespread changes in ecology. It is almost certainly unavoidable that some of those changes will be undesirable. The essential part of the problem, then, is to obtain changes for the public good at the lowest possible cost in changes for the public ill. In fact this should always be the conscious aim in every human activity that affects our environment. In the case of insect-control, that maximally desirable balance certainly has not been obtained. Its achievement must certainly involve a sort of ecological engineering that takes into account all the ramifications of biological interaction in communities. Here again, a little, narrow biological knowledge has created problems that only more and broader biological knowledge can solve.

Misguided attempts to deal with part of an ecological system while ignoring the whole are so common that examples could be endlessly multiplied. One with which I am familiar and about which I feel bitter is so-called predator control. Under that euphemistic label our tax-supported wildlife agencies are largely devoted to the extermination of wildlife. (In their jargon "control" means complete annihilation, just as the good ladies of Evanston, Illinois, mean absolute prohibition when they say "temperance.") That program, intended primarily to aid stock ranchers, has generally had little result or adverse results even for the stock. It has also set up undesirable chain reactions: fewer predators, more rabbits and rodents, so "control" the rabbits and rodents, and so on. Although some results have indeed been for the public good, on balance the most conspicuous successes have been in authorizing or providing victims for blood sports and making jobs for killers. Again, and over and over again, the fault has been attending to the part and neglecting the whole, and the remedy is in a biology that does attend to the whole.

Only in passing and to suggest to you a plethora of other examples I will mention the general deterioration of his own environment that is being caused by man. Clearing of land by lumbering and for farming has resulted in intensified erosion and floods. Plowing of good grazing but poor farming land has produced dust bowls. Mining, milling, and smelting practices have produced bare, hideous miles of piled overburden, gravel, tailings, and slag. Insecticides (again), fertilizers, detergents, industrial effluents, radioactive wastes, and sewage have made many once clear, beautiful, and productive streams and lakes nauseating and lethal to man or fish. The mining of ground water (that is, pumping it out faster than it is replenished) has produced permanent deterioration, even in areas with ample rainfall. In Florida, for instance, excess pumping has made freshwater sources turn salty. Irrigation practices in arid regions have leached soils or impregnated them with alkali, making them sterile.

The very atmosphere has been so polluted that clean, fresh air has become an ironic memory. Carbon dioxide added to the atmosphere by man since the industrial revolution and invention of internal combustion engines has radically changed the composition of the atmosphere as a whole. It is producing climatic changes with largely unpredictable and potentially disastrous results. Smog, once the prerogative of Los Angeles, now envelops all our cities and areas downwind of them up to hundreds of miles, while respiratory diseases increase enormously even beyond those due to cigarettes. Radioactive fallout, in spite of wide fluctuations, has a definite trend to increase and will increase still more now that countries without official obligations and devoid of any ethical compunctions are entering the nuclear field.

Things equally or even more dreadful are being prepared without our being more than dimly aware of them. Who knows what witches' brews of unpredictable—or, at least, certainly unpredicted—side and chain effects are seething in the cauldrons of the laboratories. We do know that vast preparations for biological warfare of unimaginable horror are going forward under a cloak of almost complete secrecy and with no check or mandate from the citizenry or its responsible representatives. The mildest to become public is the recently intensified use of herbicides to kill crops in Vietnam—an activity immoral in principle and asinine in policy, since it is sure to harm our supposed friends more than our proclaimed enemies. This operation to spread poison and destroy food in a country we pretend to be helping is called, with almost incredible cynicism, "Ranch Hand."

All of these baleful changes induced in our environment—and a great many more could be mentioned—have resulted from human activities intended to produce benefits. In every case the hope for benefit has been concentrated on one small aspect of the immediate situation and the usually maleficent results inevitable in the ecological system as a whole have been neglected, either through ignorance or deliberately for personal, local, or for national gain at the expense of others.

In many cases the benefits are real and there is no question of foregoing them. We are not going to abolish combustion engines (now including jets and rockets) because they poison the atmosphere. We can and eventually, lest we perish, we must work for a maximum ratio of benefit to harm. We must, therefore, and further know what the harm really is (or will be) and to whom it is chargeable and by what moral right. Even the beginning of advance toward that goal requires knowledge, immeasurably greater than we now have, of biology through the highest levels of the hierarchy of organization. For these human activities that affect whole nations, whole continents, and indeed the whole globe, both the acquisition and the application of such knowledge will require solidarity and co-operation on a scale that has never approached achievement, that has hardly even been hoped for in the past. What chance there now is, I leave somewhat sadly to your judgment.

There is another aspect of these literally vital problems that must be emphasized even though the mere mention of it is unpopular. That is that, at present, research and support for research over-all may be more likely to aggravate than to alleviate the sort of problems that I have exemplified. Some assistance to untrammeled research in integrated biology is provided by the National Science Foundation, the National Institutes of Health, private foundation and university endowments. Even these, however, are under pressure to restrict grants for free research and by far the greatest public support for any and all scientific research is through the Department of Defense and NASA (the National Aeronautics and Space Agency). Such backing is entirely mission-directed, that is, it is intended to further directly or indirectly the specific aims of the organizations named: for "Defense," the waging of war; for NASA, the exploration of space. Neither has any interest, or (in the jargon) has any mission, to improve the conditions of man's life here on earth, or to make the continuation of that life possible. A cynic might even maintain that this is contrary to their mission. I am not suggesting that these organizations be abolished, but only that they are of no use for solving many of our most serious problems.

Industrial research is also intensively and increasingly pursued, but we cannot look for help there either, or, at least, we are receiving none from there. This research, too, is entirely mission-directed and in an even narrower sense. In fact it is precisely this research, even though it is both necessary and useful, that has produced many of the serious misapplications of biology and drastic deteriorations of the environment. It is especially narrow industrial research that requires a counterbalance not reasonably to be expected from industry itself and at present not being provided sufficiently from anywhere.

The great hope is in the colleges and the universities, but even here there are trends that are more likely to increase than to lessen the problems of human ecology and survival. The demands of mission-directed research for missions either opposed or irrelevant to our biological needs are great, and the funds for them are liberal. It is increasingly true that the research capacity of academic biology is devoted to NASA, to Defense, or to industrial ends. Some university researchers with questionable ethics who were more or less surreptitiously working for Defense have recently been caught red-handed if not red-faced, but that does not end the matter, and most of this research imposed from outside is pointed to with pride rather than shame.

Even the residuum of research that is free, in the sense that its direction is not dictated by nonacademic agencies, has recently had tendencies that were unfortunate from the present point of view. As in other sciences, the necessities of specialization in biology have militated against the broader approaches required by such problems as I have cited. Moreover, a combination of circumstances has made the biological specialization most popular in the faculties themselves and with the less mission-directed granting agencies the one that is narrowest, least inclusive, and from the present point of view least promising of all: specialization at the subbiological, molecular level. Fortunately that is about over among the younger and the more flexible molecular biologists, who are now relating their special studies to organisms and hence to biology in a broader and truer sense. This reintegration, however, has not yet reached all administrative levels.

Now I turn if not with more optimism at least with a probably welcome change to some speculations about future biological developments affecting mankind. I mean developments other than the possible defeat of our medical aims, probable deterioration of our environment to a possible lethality, and such things as I have already mentioned or implied. I emphasize that these are, for the most part, speculations rather than predictions, because to venture firm predictions just now would take someone more daring than I, or wiser, or both. Or, now that I think of some of the predictions lately publicized, this can also be done by people a great deal more stupid than most of us.

One common prediction is nevertheless fairly certain: unless we start killing each other off still more rapidly, the earth is going to be even more crowded with people than it is now. The population explosion has other factors, for instance psychological ones, but it is largely caused by just the two main fields of applied biology:

by medicine, which has enormously increased average life spans, and by agriculture, which has enabled enormously larger populations to subsist even though, in many regions, at a near-starvation level. Speculation comes in, not with the fact that population increase is now out of hand, but with hopes or fears as to when and how it will be ended. Those few who still see no prospect or need for an end are simply foolish. The end will come, if only because there is a limit to how many people can find footing on the earth. We can only hope that the effective limit will not be imposed by starvation, recurrent epidemics, and warfare.

The only likely and humane alternative is limitation of births, not only here but everywhere on earth. There is no immediate prospect of that remedy. At present there is no considerable area where the population is not increasing at madly uncontrolled rates. This has been said so often that there is danger of its becoming merely boring, just as we have been so often warned of atomic weapons that most of us act as if they did not exist. There is, however, at least one aspect of the coming population density which, while not wholly neglected in such discussions, has been less often stressed. That is, its probable effect on natural selection in man.

The old slogan that natural selection is the survival of the fittest has led to serious misunderstanding. Since most of us think of fitness as obviously and invariably desirable, it sounds as if natural selection were always a good thing. That is not necessarily true. Natural selection has often involved changes that can hardly be considered desirable from a human point of view, for example evolution of parasites from free-living ancestors. The record of the history of living things also shows that the vast majority of species, which have arisen under the control of natural selection, have become extinct. Anomalous as it seems, the eventual outcome of the so-called "survival of the fittest" is extinction of the species. In fact what natural selection favors is always and only the genetic characteristics of those organisms that leave the most descendants under existing conditions in any one period. If those leaving the most descendants are the fittest in some sense, for example the most intelligent, then natural selection does maintain or increase fitness in that special sense. But if, as can happen and is indeed quite possibly happening among us today, those less fit in one special sense, less intelligent, say, have more descendants, then natural selection will *decrease* fitness in that sense.

Natural selection is certainly intense in the human species as a whole at the present time. With greater population density it is more likely to increase than to lessen. We know very little about the current tendency of natural selection and can only speculate about its future effects. Nevertheless, we can make some reasonable guesses about the sort of people who are likely to leave the most descendants under increasingly crowded and finally extremely needy conditions of life. It seems to me that most of those people are going to be less intelligent, less provident, and less cooperative than the average in the present population. They will increasingly be the people who can best tolerate or who even prefer conditions of population density such as exist today in our worst slums or ghettos of the underprivileged. If that is true, human nature will indeed change, and the change will entail loss of what is generally considered most refined, most civilized, most humane, and most admirably human among us now. Of course if that comes to pass our successors will not suffer from it, for they will consider much that we prize—space, fresh air, creativity, gentility, and so on—as obnoxious rather than desirable. Much that we find abhorrent, notably depersonalization and the involution of social contacts, they will prize.

The possible if not probable degenerative effects of natural selection on our species could be countered by artificial selection. The essential difference between the two is another point often misunderstood. It is not just that one is effected by nature and the other by man. Both operate by differential reproduction, but in natural selection the outcome is directly and unconditionally determined by success in reproduction. In artificial selection that success is controlled so that it favors characteristics chosen by some other criterion. There is at present no such control over human reproduction; natural selection of humans is untrammeled and is quite likely tending to reduce average intelligence, for example. But if means were found to insure that the more intelligent people on an average had more descendants, such artificial selection would halt or reverse any opposite tendency. Again I leave to your judgment what chance there is that effective and at the same time ethical controls can be applied before it is too late.

There is another alternative that has lately begun to receive considerable publicity, sometimes under the name of "genetic engineering." This is the possibility that human genetic traits can be not merely selected but made to order. There are conceivably three different approaches. One would be the induction of mutations. It is in fact simple enough to induce mutations; it is much harder to stop them. The trouble is that with trivial possible exceptions we do not know how to induce one particular mutation, and we know even less how to insure that an induced mutation will be desirable. The second possibility is that genetically defective tissues might be treated by insertion of genetic material in the form of normal DNA. For example, liver cells lacking DNA information leading to some normal molecular synthesis could be cultivated in test tubes, have the needed form of DNA inserted, and then be reimplanted in the patient's liver. Something approaching this has been done with bacteria, and its performance in humans has thus become conceivable, although the remaining difficulties are stupendous. If they do become practicable, such repairs will not be hereditary but will be a form of remedial medicine analogous to the present use of insulin to ameliorate a genetic defect resulting (under certain circumstances) in diabetes.

Most radical is a third suggestion about genetic engineering, that it may become possible to make genes to order and to insert them in human germ cells so that they would henceforth be inherited. This has had some sensational publicity and has also been viewed with alarm by those who like to view things with alarm. The excitement is premature, to say the least. Just a few of the impediments are: that we do not now know the actual structure of any human gene; that we do not know how to insert or replace genes in germ cells; that we do not know precisely how any gene produces such important traits as intelligence or temperament, or for that matter even such simple characteristics as stature; that the genetic system is an interacting whole so that insertion of a synthetic gene if it worked at all would have unforeseen and probably disastrous results. It is that last point that suggests that genetic synthesis, if possible, would be more likely to work for the public ill than the public good. I am not completely optimistic about the capacity of humans to act collectively in a rational way, but I do think that if genetic engineering produces deformed children, few individuals will have recourse to it. Anyway, the synthetic gene, if possible at all, is so remote that we had better worry about things much more imminent.

Some of my remarks on the public effects of biological science have been gloomy, and I would now like to close on a more constructive and cheerful note. This is justified and can be done by merely restating what has, I hope, been evident throughout. On the whole, the effects of biology, both scientific and technological, have been for the public good. When technological aspects have been for the public ill, their harm has rarely been absolute but has usually meant that there is a balance of good and ill and that we have paid unnecessarily much for the advantages obtained. In practically all cases that has been because a narrow, one-level approach failed to foresee and to minimize undesirable ramifications and chain effects. The remedy and, for the future, the safeguard is to work for and to insist on a more integrated biology, one embracing the whole of biology in all its aspects, from the molecule to the ecosystem. There is even evidence that biology is in fact moving in that direction, and some of the evidence comes from quite unexpected quarters, for example from laboratories supported by such suspect, mission-infatuated organizations as the Atomic Energy Commission.

CHAPTER FIVE

Science and the Culture of Our Time

"Culture: the art of developing the intellectual and moral faculties . . . ," "behavior typical of a group or class . . . ," "a particular state or stage of civilization." —Dictionary definitions

Long ago, when I was trying to learn the principles of exposition, I was persuaded that the first of those principles is to know what you are talking about. I must therefore start by deciding what I mean by "culture" and by "science." That is not a simple matter. Many ponderous books and innumerable essays have been devoted to the definitions of those two words. That definitions do not agree and that so much effort has been spent on them already tells us something about culture and about science.

The idea of culture that first comes to mind for many people might be labeled as the "braves and squaws" or perhaps the "Jiggs and Maggie" point of view. Culture is what women do or are interested in. That idea finds considerable support in a recent sample poll of supposedly cultural interests of men and women. It was found that a major cultural achievement of women, but not of men, was to have read a book all the way through. Just half the women, but many fewer men, had accomplished that feat at one time or another in their lives. Since many people outside of colleges do not distinguish between books and magazines, it is probable that the "books" that so many women had read all the way through (including, no doubt, the advertisements) were movie fan magazines.

On the side of Jiggs or the braves, a major cultural accomplishment of men was to have gone to a football game. They were well ahead of the women there. The men were also ahead in watching TV westerns, but in this respect the women were not far behind. As far as this poll reveals, watching TV westerns is the predominant cultural interest for both men and women.

Another view of culture not entirely distinct from the Jiggs and Maggie school is that of the robber barons. Since the poor are uncultured by definition, those who rose from rags to riches felt it necessary to demonstrate that they had acquired culture along with their dollars. Or perhaps their wives felt it necessary. In any case, that seems to have been the motive for competitive buying of the works of long-dead painters, for support of grand opera (to this day a motif of the Jiggs and Maggie epic), and for the founding of eponymous universities. I am not poking fun or, at least, if I am, it is in an indulgent way. I believe that the robber baron theory of culture did as much good as harm.

A still older view of culture, and one that certainly has not done as much good as harm, is that of the Greeks and the barbarians. I refer to the ancient Greeks and their neighbors, but the attitude is faithfully preserved in diluted form in the microcosms of many college campuses. The principle involved here is that what "we" do is cultured and what "they" do is not. It is, of course, implied that our culture is in all respects better than their nonculture. One may supply any desired antecedents for "we" and "they." Obviously everyone subscribing to this view has his own definitions. We know what came of Hitler's conviction that Nazis were cultured and Jews were not—and this, incidentally, illustrates the moral view that the uncultured have no rights, not even the right to live. Greeks did not hesitate to enslave barbarians or Nazis to exterminate Jews. Closer to home, you know—although I am sure you do not approve—the application in which people with light skins become "we," the cultured, and those with dark skins are "they," the uncultured without inherent rights.

Let us spend no more time on such foibles. Yet the time spent has not been wasted or beside the point. Those attitudes toward culture bear on our topic just because they are so utterly unscientific in basis and because they exclude science from the supposedly cultural aspects of life.

That exclusion is also commonly made by a verbal dichotomy which, if also illogical, is at least more innocuous. We are all accustomed to the academic contrast of the humanities and the sciences, even if we are often uncertain as to where the line should be drawn. I sought guidance in a dictionary of high reputation and there was told that the humanities are "the branches of polite learning." The editors must subscribe to the Greek versus barbarian view of culture. The humanists are the polite Greeks and the scientists are the impolite barbarians.

The division of learning into humanities and sciences is not the illogicality to which I referred. What is illogical is implied in that dictionary definition: an equation of culture with the humanities. The sciences, impolite, uncultured, should perhaps be labeled as the inhumanities. That—alas!—is not a facetious remark. It fairly represents the attitude of many humanists, such as Jacques Barzun, to name only one as evidence that I am not inventing. (By "humanist" I here mean one who pursues the academic subjects called humanities; the word "humanist" also has other honorable but almost antithetical connotations, such as one whose ethical or religious outlook centers on the worthiness and welfare of human beings.)

Since I am myself a scientist, I prefer a more scientific attitude toward culture. It occasionally is tempting but it would not be scientific to turn the tables and to argue, as one really could with considerable evidence, that "we," the scientists, are cultured and "they," the humanists, are not. For the definition of culture that we want, we can best go to the scientists whose business it is to study culture or, rather, cultures in the plural. With variations in the wording, most of them agree to define culture as the sum total of what the members of a human social group believe, make, use, and do beyond what is genetically fixed. In simpler terms, culture includes human artifacts and learning. Still more simply, culture is just the way people live.

A virtue of those definitions and their value for our topic are that they make it plain that science is a part of culture. Under them we can abandon silly attempts to contrast science and culture or to weigh one against the other. They permit us to consider the real and important point: what role does and should science play in our culture?

Definition of science, like definition of culture, requires first of all the discarding of some popular concepts, not necessarily complete mistakes but at best not what I shall mean when I speak further of science. It is fairly certain that most people if asked for an example of science would refer first to the space program, probably second to atomic or nuclear weapons, and third to some commercial goods, household gadgets, perhaps TV, perhaps at a slightly more sophisticated level an infrared rotisserie, or perhaps at a decidedly more vulgar level a new brand of toothpaste. It is true that all of these have incidentally, one might almost say accidentally, made some slight contributions to science. That is part of what the defenders of the space and weapons programs call fallout—a word with most unpleasant overtones. On the whole, however, the three tremendous fields to which I have referred are consumers and not producers of science. They produce little or no new or fundamental knowledge about the world we live in, but they use such knowledge for technological ends.

It has often been insisted that there is no clear line between science and technology. That is true, but the conclusion sometimes drawn that therefore science and technology cannot or should not be distinguished is nonsense. In spite of marginal cases, science and technology are typically quite different, and they have decidedly different roles in our culture. The man who designs a rocket, a warhead, or a refrigerator is a technologist, not a scientist, and his cultural function is definitely nonscientific. That does not necessarily mean that his role is less important than that of a scientist—although to be frank with you, I believe that it is. The point here is simply that his activities and his functional situation in society are not at all the same as those of a scientist.

Technology and even engineering are often called science or, if any distinction is sought, applied science. It is not depreciating those essential activities to insist that they are not, in fact, science at all. Science, strictly speaking, is what technologically oriented people are wont to call basic science or pure science. At least I shall use it only in that sense. Among many possible definitions of science, I prefer the following: science is an exploration of the material universe that seeks natural, orderly relationships among observed phenomena and that is self-testing.

The cultural implications of that definition of science will be explored shortly. It should here be emphasized that most of the voluminous discussion of science and the culture of our time has either failed to distinguish between science and technology or has referred primarily to technology rather than to science. Perhaps I may return to the example of Barzun without offense, accepting him as a brilliant humanist who is in some way a spokesman for humanistic attitudes toward science. He does distinguish science from technology, which he calls *techne* for reasons of subtle distinction that I need not pursue. He says in so many words that "the life man has made for himself is not worth living"—although I am happy to note that he continues to live it. The principal reason that he gives for his distaste for life is that it is lived within a scientific culture. (He also dislikes modern poetry, modern art, modern teaching except his own, and many other aspects of modern culture, but that is beside our present point.) What he specifically objects to in the "scientific culture" is in almost every instance technology, and not science at all. In fact, he admits by implication that science might be rather nice if it could be kept in its place. Thermodynamics could at least be amusing but automobiles and refrigerators are disgusting. (Oddly enough, I do know humanists who ride in automobiles and preserve food in refrigerators.)

The faults and virtues of technological products and developments have been endlessly argued and will surely continue to be. That dialogue is both interesting and important, but I do not think it needs another contribution at this point, at least not one from me. My topic is indeed *science,* and not technology, in our culture. One more point of confusion between the two must be cleared up: the belief that the United States is remarkable in the extent of income and resources devoted to science. That is true to some extent, but is not as extensive as most people seem to think.

Data on the support of science usually are included in figures for research and development—now widely called "R & D." The criteria for science, strictly speaking, and for the technological and engineering aspects of R & D are equivocal and various. It is impossible in almost all cases to determine precisely how much of reported expenditure for R & D is for science. The federal budget for fiscal year 1966 included approximately $15.5 billions for R & D in various departments and agencies. Only about one-seventh of that sum was for nominal "basic research," which includes what I am calling science but also some activities that I would call technology. Somewhat older and less accurate figures for total spending in the United States and not only the federal budget indicate unquestionably that less than one-tenth of total R & D spending is for science. A lower proportion than for federal funds is of course to be expected, since industrial R & D is here included, and extremely little of that is for science in the strictest sense.

Alan Waterman, former director of the National Science Foundation, has pointed out another important distinction concealed within the global figures for R & D. Research may be "mission-directed" or "free." Mission-directed research, even if in so-called basic science, is expected to produce technological results in a specified field. The nature of this research is determined by the management that provides the funds and not by scientists as such or by the needs or interests of free scientific inquiry. Its *purpose,* even if not its substance, is technological. The greatest part of federal support for science (well over half) is administered by the Department of Defense and the National Aeronautics and Space Administration. One can take it that virtually all research supported by those agencies is mission-directed. Waterman estimates that about 80 per cent of federal support for "basic research" is mission-directed. Since, as previously noted, what is called "basic research" receives about one-seventh of federal R & D-budgeted funds, this means that *only about one-thirty-fifth* of those funds is for *free* basic research, in other words, for the advancement of *science* in its own name and right. It sounds grand that the total federal budget for, as is sometimes said, "science"—actually for R & D—is some $15.5 billions. But of that almost certainly less than $500,000,000 is for free research. That is, to be sure, a nice round sum, but not so nice or so round as $15.5 billions. I have no sufficiently up-to-date and accurate estimate for free research on nonfederal funds, but the percentage of total R & D funds is not likely to be more. For instance, it must be assumed that practically all industrial research is mission-directed. I would agree that industrial research *should* be mission-directed, but that is a good reason why more nonindustrial research should *not* be mission-directed.

A point here is that expenditure for science in the United States is practically insignificant from a direct economic point of view, in terms of gross national product, total tax revenue, or any other pertinent yardstick. Expenditure for technology and engineering, even more specifically for space and defense, *is* significant in that sense. Those activities are far from depending totally on underlying science, as some have maintained, but they do have considerable dependence. Indirectly through them science is indeed an economic force, but that is quite distinct from its other cultural roles, to which we may now turn more explicitly.

In preparation for more direct consideration of what it means for our culture, let me repeat my previous definition of science: science is an exploration of the material universe that seeks natural, orderly relationships among observed phenomena and that is self-testing. The most important implication of that definition is that the phenomena of the universe do have natural and orderly relationships. A scientist assumes that the universe makes sense. My attitude toward the ancient Greeks shocks historians and some historically minded scientists. I do not think that any of the ancient Greeks, not even Aristotle, deserve to be called scientists. Some of them were marvelous architects and sculptors. Few have ever exceeded their best poets and dramatists in understanding of the human heart in weakness and in strength. But their so-called science was for the most part childish or ridiculous or both. Yet some of their philosophers, including Aristotle, did assume that the world makes sense, and that respectably antique notion was still around when the first real scientists began to appear in history, some eighteen centuries after Aristotle.

Among the ancient Greeks the belief that the world makes sense seems to have been confined to a handful of philosophers and never to have characterized the culture as a whole. Its spreading cultural influence began with the Renaissance in western Europe. By now it has affected all cultures, everywhere on earth, but in greatly varying degree. There are still tribal Indians in South America who have never seen a European, and we can assume that they are not much affected by the scientific attitude. We consider ourselves to be most strongly affected, and that is true in a comparative sense, but I shall have to point out that we are still far from a complete acceptance of an orderly universe.

To that basic concept of orderliness, a further idea was rather slowly added in science: that phenomena can be understood by rational investigation, and assurance gained by induction from observations and by predictive testing. Those concepts, incidentally, were almost entirely foreign to the Greek spirit of deductive philosophy. When scientists had finally reached that point, they no longer believed that an odd or hitherto unexplained phenomenon represented an inexplicable, or supernatural, or simply disorderly element in the universe. They sought rational, natural explanations, with varying but ever increasing success. Slowly and often in the face of bitter opposition, this attitude began to take a real, even if always limited, hold on nonscientists. It began to be a significant element in European culture and to spread thence into other cultures.

Even while still inchoate, this change in culture brought to the human spirit liberation from needless apprehension and gross superstition. Comets had been dreadful and awful omens, in the old, undiluted senses of dread and awe. They struck mankind with horror. As late as 1755 it was seriously supposed by pious and, perhaps, otherwise decent people that the tens of thousands killed in an earthquake at Lisbon had received justly deserved punishment from a vengeful God. Now comets make such simple good sense that the recurrent ones can be adequately predicted. We are not yet very good at predicting earthquakes or even daily weather, but we can usually explain them after the fact, and we are confident that the failures of predictions are not because the phenomena are unnatural or disorderly but just because they are too complicated for us. These changes in the intellectual climate of culture, limited as they still are, have removed a tremendous burden from mankind.

It is through what I have called the climate of culture that the effects of science have been greatest. Since antiquity it has often been assumed that the norms of culture represent human nature, and that human nature does not change. It has also been claimed for a shorter time that science is simply systematized or glorified common sense. Both statements are untrue, indeed obviously so. The norms of culture differ greatly from one culture to another, even in our present relatively homogenized world. They have changed greatly in history. If these norms have any relationship to human nature, then human nature does change and quite radically. Since science was invented, it has become the most important agency of such change—clearly more important than technology, for instance, which existed long before science.

As for common sense, if the expression means anything (a moot point), it must mean the ideas usually held by rational people. But if such be common sense, then science has more often been a contradiction than a confirmation or mere refinement of common sense. To our forebears and even to a few people today there could be no more obvious common sense than that the earth is flat and the sun moves over it daily. To us it seems merely common sense that the earth is round, that it rotates daily, and that it revolves annually around the sun. But there is here nothing obvious or objectively observed. It is all due to science; it is in truth a scientific theory and not an observed fact. Science has created "common sense" and a new climate of opinion.

Organic evolution is just such a theory as the theory of the solar system, except that there is a great deal more evidence for evolution. Special creation was once common sense, but now it is a discarded dogma and evolution is common sense. To be sure, there are very few people who disbelieve in the theory of the solar system and many more who disbelieve in the theory of organic evolution. That does not indicate that evolution is not now common sense, but just that certain people can have common sense about some things and not about others.

Another interesting example further illustrates the shift induced by science in a cultural climate and in what is common sense. Voltaire, hardly a scientist himself but quite familiar with the science of his time, weighed many substances before and after burning them in an endeavor to learn the weight of fire. Some of the burned substances gained weight, some lost it, and some remained the same. Voltaire remained baffled! In his day it was scientific lore and also common sense either that fire is an element, as the ancient Greeks "knew," or a material called "phlogiston." Now everybody knows—it is common sense!—that fire, or better, combustion is a process, not an element or substance, and that poor old Voltaire was merely ridiculous for trying to weigh a process. Yet we can ill afford to vaunt our common sense over his. Our wise men still debate whether mind and body (not to mention soul and body) are two different "things" or two aspects of the same "thing." I have not found many who see, what is surely the rising common sense, that this is quite analogous to the old mistake about phlogiston. Mind is not a thing or an aspect of a thing; it is a process that occurs in the body. We should no more argue about the distinction of mind and body than about the distinction of, say, digestion and body or respiration and body.

Science is not only something that has altered common sense and the cultural climate. This is a mutual influence, a two-way street. Science not only affects but is also affected by the culture in general. During the Middle Ages there was no lack of brilliant and learned men, but the climate of western European culture was then unfavorable to concepts necessary for science to arise. That climate changed in the Renaissance in just the way necessary for the invention of science. Even countries such as China, with more ancient and more complex civilizations than Europe, did not invent science, because their cultures were opposed. Only recently have such cultures changed enough for the importation of scientific knowledge and ideas. Organic evolution and even natural selection were hypotheses well known before Darwin, but only after about 1850 was the European cultural climate such that acceptance of those ideas could begin. Mendel's discoveries in genetics were fallow for thirty-five years not because they were unknown but because the climate of the culture was not ready for them in 1865.

The cultural climate of the times has not only had an inhibiting effect as in those examples; for better or for worse, it has also been a stimulus. Atomic physics would certainly not have progressed so rapidly in recent years if our cultural climate had not included an intense interest in killing our fellow men. On the other hand, that catastrophic development, in combination with a multitude of other factors, has also stimulated interest in more peaceful sciences and technologies.

Now European and American cultures are impregnated with science to an extent never equaled by other or earlier cultures. Even comparatively trivial parts of our lives—perhaps especially these—are affected by our confidence in a sensible universe, in objective approaches to problems, and in their rational and logical consideration. A true and pertinent example that nevertheless verges on the ridiculous is provided by gamblers who follow what they call "scientific" systems. The bettors never have a really scientific system (it is the bookies and the house that do), but they would not consider it scientific, or even have a system, if science had not become so intimate a part of our culture. Less trivially, we face most of the decisions of life with the intention, at least, of obtaining the relevant facts and acting rationally on them: what to buy, where to go, what to learn, what to think of or to do about drought or fire, and innumerable other decisions. In most situations we try to use common sense, and I think it is now evident enough that the common sense of our culture is strongly affected by scientific concepts and different from common sense in nonscientific communities or prescientific times.

There are, nevertheless, decided limitations in our culture's adoption of scientific attitudes. Ever since science arose, the most obvious point of stress has been between science and religion. Conflict has arisen whenever theology has taken a dogmatic attitude on objective phenomena of the material universe. Such phenomena are open to independent investigation by scientific methods, and if the results disagree with dogma, the dogma becomes rationally untenable. A religion that dogmatically demands denial of organic evolution, for example, is quite simply a false religion. In fact, the religions that give any value to rationality eventually have retreated from tenets, if not technically from dogmas, when these conflicted with clear scientific findings. Even religions that claim possession of absolute, immutable truth have in fact changed notably under the impact of science. That is another way in which science has greatly affected the culture of our times. The compatibility of science and religion—and I believe that they can be compatible—depends on recognition that each has a sphere in which the other is inappropriate.

Here I do not want to follow up that topic, but rather to consider cultural failures to adopt scientific attitudes where those attitudes are unquestionably appropriate and desirable. Unfortunately, such failures are widespread and dangerous. Many examples are familiar to everyone.

I cannot help daily sight of one, for my morning paper has an astrological column aptly but inescapably next to the comic strips that I read each morning. But not so apt, after all, for there is nothing comic about widespread present-day acceptance of a superstition from which the last vestige of rational support was stripped by Newton almost three centuries ago. Or if I pick up a magazine, there is likely to be an ad for "The Brandy of Napoleon" or for a brand of "London Gin," the former with a hideous chromo of Napoleon crossing the Alps and the latter with a picture of the Houses of Parliament. In fact the brandy has no connection at all with Napoleon Bonaparte and the gin is made in New Jersey. Perhaps that is not so much a failure of scientific attitude as a simple demonstration that some advertising men—I have not said all—are liars or morons or both. The latter possibility is confirmed by a recent news story of an advertising executive who testified that he was so impressed by his own manifestly false claims for some pills that he started to take them himself!

The most distressing and also among the most common examples of cultural failure are provided by advertisements not so simple-mindedly calculated to deceive but using a pseudoscientific approach. Here at short intervals between bits of an entertaining TV show is a commercial for the tablet that has "more of the ingredient that doctors recommend." Does it never occur to anyone that the tablet with *most* of the ingredient that doctors recommend would be made entirely of that ingredient? The voice of God in the commercial of course does not name the ingredient, but it is evidently plain, cheap aspirin. Next is a commercial in which the offscreen voice of God, bellowing from an echo chamber, says, "Here's *proof!*" The efficacy of a patent medicine is then shown by a cartoon of leaking plumbing inside the silhouette of a man. *That's* proof?

You have all seen those commercials and many more like them. Their importance is that they are not just false and moronic. They are an extended, continual, day and night series of lessons in illogical thought patterns. They are training young and old *not* to think rationally. There is plenty of evidence from other fields that the ability to think irrationally is indeed widespread and, I would judge, still spreading. For instance, there is the argument by "sports lovers" that guns should not be licensed because they kill many fewer people than cars do. The fallacies should not have to be pointed out, but apparently must be since the argument is endlessly repeated: (*a*) guns are made for the purpose of killing, cars are not; (*b*) nevertheless, cars *are* licensed!

If an example at a more exalted level is wanted, note President Johnson's argument for his school bill that we spend $1,800 a year on a delinquent child in a "detention home" (*i.e.*, juvenile jail) but only $450 a year on a child in public schools. The fallacies are a little more subtle here, but they are not less pervasive. The comparison is not of the same things: total support for the delinquent; no support (as such) for the school child. Even if the same things had been compared, it is not clear that spending more on the school child means (as implied) having fewer delinquents. Might we not really need more for work with delinquents, and perhaps even less for nondelinquents? (Before I am berated with equal illogic, I hasten to say that I am in favor of education and motherhood, opposed to delinquency and man-eating sharks, and, on balance, in favor of Johnson's imperfect school bill.)

Scientists are not immune. They, too, can learn to be irrational. For example, it is now just as sure as the sun will rise tomorrow that tobacco is a major causal agent in cancer and some other diseases and that it is a principal cause of unnecessarily early death in our culture. Yet in England, R. A. Fisher, one of the greatest biometricians who ever lived, went to his own death denying this undeniable fact. It is also currently denied by a number of undoubtedly competent statisticians, biologists, and physicians in this country. Here are clues to this, at first sight, incredible phenomenon: (*a*) the scientists involved are a very small minority in their respective professions; (*b*) most of them (and this included Fisher) are employed by or otherwise receive material benefits from the tobacco industry.

I am not accusing those scientists of being dishonest. It would be far better if they were, for dishonesty in science does not pay and eventually always defeats its own ends. No, it is simply that human beings are capable of deceiving themselves, of becoming irrational when personal interests or advantages are at stake. Science is by definition *capable* of self-correction, but, as Anne Roe has shown, human bias is not necessarily self-correcting even in science. This is an element in our imperfectly scientific culture that we must recognize and work against.

Another example is not so clear-cut as that of the unnecessary mortality from tobacco. There is no objective, clear evidence whatever of the existence of life anywhere in the universe except here on earth, yet many astronomers, physicists, chemists, and some biologists assert either that extraterrestrial life definitely does exist, or at least that its existence is so probable as to warrant proceeding on that assumption. In consideration of the harmfulness of tobacco there is conclusive positive scientific evidence. In the case of extraterrestrial life, what real evidence there is pertains only to our own solar system and it is strongly, although not quite conclusively, opposed to the presence of life on any other planet of this system. When scientists argue for the existence of such life, they are thus discussing a partly scientific question but are on the wrong side. Outside of our solar system there is absolutely no objective evidence at all, either pro or con. When scientists discuss that issue they are at best extrapolating probabilities beyond any presently testable point. At worst, and usually, they are simply fantasying.

My own extrapolation is that while life probably does exist *somewhere* outside of our solar system, that is meaningless for us here on earth because there is no likelihood that we will ever know for sure and the chances of usefully communicating with intelligent beings anywhere else in the universe are effectively nil. Of course it is naughty of some scientists not to agree with me, but that is unimportant. What is important is that so many of them are not behaving in a scientific and rational manner. The most positive and unsupported opinions about extraterrestrial life have been expressed by physical scientists (astronomers, physicists, chemists) who are, to put it mildly, inexpert in the pertinent subjects of biology and evolution. To put it less mildly, they do not know what they are talking about. Some biologists do go along. It is my impression that they are a fairly large minority of biologists, but there has been no poll and I may be mistaken: they may be only a vociferous small minority or may be a majority. Many of them—I think almost certainly most of them—are profiting in some way from the space industry, just as most biologists questioning the lethality of smoking are profiting from the tobacco industry.

W. J. Luyten has been quite blunt about this development in asking "Is it simply that the line for bigger NASA contracts forms to the right—in front of the rainbow labeled Life Outside the Earth?" NASA has already been mentioned as by far the largest source of nonmilitary contracts, grants, and fellowships for research. I do not agree with a possible implication of dishonesty, but as in the previous example I do believe, indeed I think it quite obvious, that self-interest has here biased judgment.

There are also at least two other factors in this situation. For one thing, it is probable that most people aware of the possibility hope that there is life, and by great preference intelligent, articulate life, elsewhere in the universe. For example, a recent book by a newspaper reporter has become an unexpected best seller because it exploits this wishful thinking. The scientific community is, I think, more resistant than most, but it is not entirely immune to any of the frailties of humanity, including the unreason of wishful thinking. Another factor is that some biologists argue that since the space program is going forward, for better or for worse, it is only sensible to take advantage of it for what marginal benefits may accrue to biological science. At first sight this attitude does seem like common sense, of a sort, but at second look it is hard to distinguish it from cynical opportunism.

Decisions that greatly affect the complexion and direction of science in our culture are not, as a rule, made by scientists or on the basis of a balanced scientific program. A NASA official recently lashed out at scientists who had criticized NASA, counterattacking those critics on the ground, among others, that some scientists had opposed programs that turned out to be feasible. That is a very interesting remark. No attention is paid to scientists' opinions as to whether a program is advisable or desirable, only as to whether it is feasible. And it should occur to everyone that this is really an engineering decision, not a scientific one, anyway.

The founding fathers of this republic were extraordinarily rational men, more so than almost any of their successors in politics. They assumed that the welfare of the country would automatically proceed hand in hand with the cultivation of science and of the useful arts (that is, technology and engineering). They expected that political judgment would arise from the accumulation of knowledge in a free society. Through the years since 1776, that view has been not only modified but also virtually reversed. Later and now largely current theories of government hold that policies and value judgments are basically political in nature and that administrators, including scientists when they are involved at all, are not to make or to evaluate decisions but only to implement them. That development has been well discussed by D. K. Price, Dean of the Graduate School of Public Administration at Harvard, who concludes that scientific inquiry is our only reliable source of factual information, but that science and scientists do not supply either an evaluation of those facts or a program arising from them. This is indeed the present situation, but I do not think that it needs to be or should be.

In this cultural climate it is often stated, not seldom by scientists themselves, that the opinion of a scientist outside his own narrowest field of technical specialization is no better than that of any other citizen. In fact, it is sometimes implied that a specialist outside his specialty is less reliable than most people. Robert Hutchins is credited with the appalling statement that, "the narrower the field in which a man must tell the truth, the wider is the area in which he is free to lie." At least the specialized scientist has *some* field in which he must tell the truth, and it is even possible that telling the truth can become habitual. The generalized humanist, as it would appear from Hutchins' wisecrack, is free to lie in any field at all.

I believe that many scientists have become too timid, even apologetic, as to the potentialities of science, perhaps in overcompensation for the excessive confidence of earlier days. For one thing—a very important thing—it is not true that science cannot deal with values or that it does not evaluate the facts that it discovers. In fact, scientists deal with values all their lives and in every aspect of their work. They become so expert at it that they are hardly aware that they are doing it, but they are professional valuers to greater extent than members of any other profession. It is also untrue that there is any objective aspect of life or of culture to which science is inapplicable. I am not saying that there are no aspects of the universe or of humanity that are not within the field of science, but only that there are no phenomena that cannot be studied and evaluated scientifically, including, for example, the humanities or theology. The truth or falsity of, say, the immanence of God cannot be examined scientifically, because there are no objective phenomena to start with and no means of self-correction if an error is made. This is not a factual question in the scientific sense. But that some men believe in the immanence of God, draw certain conclusions from their beliefs, and act in certain ways on those conclusions are observable facts and they can be not only studied but also evaluated by scientists.

I also object strenuously to the folklore that scientists are not competent, as such, to draw up programs based on the data and theories of science. Scientists are certainly subject to human failings, as I have shown. However, what reliable, factual, objective knowledge we have is a product of science, and our cultural aptitude for rationality—as far as it goes—has been developed by science. At the very least, scientists are likely to understand their data better than other people, and more likely to act rationally on that basis. You and I will not live to see it, but perhaps the dream of our founding fathers will sometime come true, and it will be the function of politicians to implement the programs of scientists. It may even happen that a scientific, that is to say a rational, attitude will pervade all aspects of our culture and that we will have not only weapons (if we must), rockets, and gadgets, but also health, sanity, and wisdom.

CHAPTER SIX

The Biological Nature of Man

> "It has often and confidently been asserted, that man's origin can never be known: but ignorance more frequently begets confidence than does knowledge: it is those who know little, and not those who know much, who so positively assert that this or that problem will never be solved by science."
>
> —CHARLES DARWIN

Those words were written nearly one hundred years ago and were published in 1871 in the introduction to Darwin's book on *The Descent of Man.* In his even better known work on *The Origin of Species*, which had appeared twelve years earlier, he had been content to say (somewhat coyly) that by that work "light would be thrown on the origin of man and his history." Others soon indicated the nature of that light. Thomas Henry Huxley's classic *Man's Place in Nature* was published in 1863, and by 1871 numerous other naturalists of the first rank had already accepted the evolutionary origin of the human species. Darwin's own contribution to the problem of man's origin firmly established two points: first, *Homo sapiens*, like all other organisms, has evolved from prior, extremely different species by natural means and under the directive influence of natural selection; and second, man is the descendant of apes or monkeys of the Old World.

Darwin's first point, that man is the product of evolution involving natural selection, has been attacked on emotional grounds, but it was not and is not now honestly questionable on strictly scientific grounds and by anyone really familiar with the facts. The second point, of man's descent from an Old World ape or monkey, was for some time more open to scientific dispute. However, here, too, the debate was often more emotional than objective. In some pedagogic circles it became usual to maintain that man is not descended from an ape but from a common ancestor neither man nor ape nor, if one cared to go still further afield, monkey. Some went so far as to attempt to enlist Darwin posthumously in their own pussyfooting ranks by saying that he never maintained that man arose from an ape but only from a common ancestor . . . and so forth. In fact, although Darwin was slow to enter the dispute, when he did so he was more honest than those supposed defenders. He flatly said, "We must conclude, however much the conclusion may revolt our pride, that our early progenitors would have been properly . . . designated [as apes or monkeys]." The unscientific and really uncalled-for remark on pride does little to modify the forthrightness of the conclusion.

Darwin's conclusions in 1871 already covered what is most vital for consideration of man's biological status. Subsequent discovery and study have fully corroborated Darwin and have added an enormous amount of detail. That is interesting and important, and most of what I have to say here concerns it. At this point, however, the essential thing is that Darwin put the whole subject of the nature of man on a new and sound footing. To be sure, in the introduction of *The Descent of Man*, from which I have already quoted, Darwin went on to say that, "The conclusion that man is the codescendant with other species of some ancient, lower, and extinct form, is not in any degree new." He then cited Lamarck, Wallace, Huxley, Lyell, Vogt, Lubbock, Büchner, Rolle, Haeckel, Canestrini, and Barrago as "having taken the same side of the question." In fact, as regards this particular point, Darwin was doing too much honor to those worthies, some still famous and some now forgotten. It is true that they had all discussed the descent of man before Darwin himself did so in an explicit way, but with the sole exception of Lamarck they had done so after publication of *The Origin of Species* and on the basis of that work by Darwin. As for the few who really had postulated an evolutionary origin for man before *The Origin of Species*, their views were largely philosophical speculations inadequately or not at all supported by objective evidence and sometimes, as in the case of Lamarck, reaching a conclusion only approximately correct on grounds that were flatly wrong.

The question "What is man?" is probably the most profound that can be asked by man. It has always been central to any system of philosophy or of theology. We know that it was being asked by the most learned humans 2,000 years ago, and it is just possible that it was being asked by the most brilliant australopithecines 2,000,000 years ago. The point I want to make now is that all attempts to answer that question before 1859 are worthless and that we will be better off if we ignore them completely. The reason is that no answer had a solid, objective base until it was recognized that man is the product of evolution from primeval apes and before that through billions of years of gradual but protean change from some spontaneously, that is, naturally, generated primordial monad.

It is the biological nature of man, both in his evolutionary history and in his present condition, that presents us with our only fixed point of departure. These are the facts we can find out for ourselves, in great, ever-increasing detail and soundness, open to all of us in irrefutable observations. Their interpretation is in some respects ambiguous and disputable, but interpretation at a given point becomes increasingly clear and undisputed as time goes on. Doubtfulness moves outward with the expanding frontier of knowledge.

I do not mean to say that the biological study of man or even that the scientific study of man in terms broader than biological can here and now—if ever—provide a satisfactorily complete answer to the question "What is man?" The other, older approaches through metaphysics, theology, art, and other nonbiological, nonscientific fields can still contribute, or can now contribute anew. But unless they accept, by specification or by implication, the nature of man as a biological organism, they are merely fictional fancies or falsities, however interesting they may be in those nonfactual categories. I am here concerned with man's biological nature in a rather broad sense, on the ground that this is a necessary, even though it is not a completely sufficient, approach to comprehension of man's nature.

Already in Darwin's day it was clearly established that among living animals the great apes are anatomically most similar to man. Some anatomists, reluctant to acknowledge their poor relatives, stressed differences between man and any apes: the larger human brain, obviously; the longer and less divergent first toe of man; the absence or, more commonly, the only sporadic presence in us of certain apish muscles and other structures. Such discussions completely missed the point. Of course men and apes differ. In itself, that means only that we belong to different species. The point at issue is not whether we differ, but in what way and how closely the different species are related.

All later study has corroborated the special relationship between men and apes and has made knowledge of it more precise. The evidence has lately been greatly increased in extent, in detail, and in its basic character. It now includes such fundamental points as the numbers and shapes of chromosomes, the exact molecular structure of hemoglobins, the resemblances and differences of serum proteins, and many others. All the evidence agrees and the conclusion is unequivocal. Man is not identical with apes in these or other respects. However, he is clearly related to the apes, and among the apes he is most particularly related to chimpanzees and gorillas, which are closely related between themselves. A necessary inference from this evidence is that the common ancestor of apes and men was itself a member of the ape family. Not only that; we had a common ancestor with gorilla and chimpanzee after their ancestry had become distinct from that of the other living apes (orangutans and gibbons). Our relationships to gorilla and to chimpanzee are about equal, although gorillas may have become somewhat more specialized with respect to the common ancestry.

More precise evidence as to relationships and as to the course of anatomical change in the human ancestry must come from fossils. There are special reasons why pertinent fossils are comparatively uncommon: crucial stages apparently occurred in the tropics, where preservation and discovery of fossils are difficult and where exploration has generally lagged; populations of apes and of prehumans were always small, not at all comparable with the great herds of grazing animals, for example, common as fossils; and the habits and abilities of apes and prehumans were such as to reduce chances of natural burial and preservation as fossils.

Nevertheless, a great many fossils have been recovered and discovery is active at present. We are far from having the whole story, but parts of it are increasingly clear.

In Darwin's time only one really distinctive kind of fossil ape (*Dryopithecus*) and only one really distinctive kind of fossil man (Neanderthal) were known. From the former, Darwin correctly inferred that by late Miocene, at least, the lineages of apes and monkeys had separated. He was not clear as to the possible implications for separation of the strictly human lineage, which he thought might have occurred much earlier. As regards Neanderthal man Darwin could only express surprise that in spite of their antiquity the Neanderthals had brain capacities probably greater than the average for modern man.

Now it is known that apes more or less similar to *Dryopithecus* were widespread and, as apes go, numerous through the Miocene and Pliocene of Europe, Asia, and Africa. Present estimates place the beginning of the Miocene at approximately 25,000,000 years ago. The divergence of apes and Old World monkeys is thus at least that old. There is, in fact, some evidence that this divergence occurred in the Oligocene, which preceded the Miocene and began some 10,000,000 years earlier. Divergence of apes and monkeys was identical with divergence of the human ancestry and monkeys, because the earliest apes were also ancestral to man. The time of the final split of the specifically prehuman lineage from that leading to gorilla and chimpanzee has not yet been closely determined. On present evidence it seems most likely to have occurred during the Miocene, that is, quite roughly between 10,000,000 and 25,000,000 years ago. The earliest known forms that may be definitely on a prehuman line as distinct from a pre-gorilla-chimpanzee line are *Ramapithecus* from India and the closely similar, indeed probably identical, supposed genus *Kenyapithecus* from Africa. Unfortunately, those animals are known only from teeth and fragments of jaws, so that their affinities are somewhat uncertain and the anatomy of their skulls and skeletons is entirely unknown. The known specimens are approximately 10,000,000 years old, give or take a few million.

The next significant group of fossils is that of the australopithecines, literally "southern monkeys" although they almost certainly were not exclusively southern and with complete certainty were not monkeys. They were surely and comparatively well known from East and South Africa, doubtfully and, at best, poorly known from elsewhere in Africa and from Eurasia. In Africa they are clearly divisible into two distinct groups. There is dispute as to whether those groups should not be subdivided still further and whether they should be called species or genera. Although the specialists can become enraged over those questions, they have no real importance for others, the important fact being simply that the two separate groups did exist, a point on which even the specialists now agree. Both groups resemble apes much more than we do now, but both are more nearly related to us than to the apes—another point on which the specialists have finally agreed after years of wrangling. They definitely belong to the human family, Hominidae.

One group, typified by *Paranthropus robustus*, retained some particularly primitive (more or less apelike) features and yet became somewhat aberrantly specialized. It cannot have been directly ancestral to modern man. The other group, typified by *Australopithecus africanus*, although also primitive within the human family, more closely resembles our own genus, *Homo*. Both groups are now believed to have appeared at least 2,000,000 years ago. For a long time, perhaps 1,500,000 years, there were at least two distinct lineages of the human family living in Africa and probably throughout the warmer parts of the Old World. One, more primitive and aberrant, showed little progress and finally became extinct. The other, more progressive, evolved into *Homo*. A matter still under sharp dispute is whether the latter lineage included *Australopithecus africanus* as our direct ancestor, or whether for a time there were not actually three distinct lines: the two kinds of australopithecines and still another more directly related to *Homo*. The latter suggestion arises from Leakey's discovery of what he calls *Homo habilis*. However, some authorities believe that supposed species not to be on a distinct lineage but to belong to the line leading from *Australopithecus africanus* eventually to *Homo sapiens*.

That dispute is interesting and we hope it may soon be settled, but it is far less important than the fact that our ancestry passed through a stage closely similar to *Australopithecus africanus* if it was not that group itself. Our ancestors were then fully bipedal, ground-living animals, using their hands for manipulation as we do but perhaps not quite so skillfully. Their teeth were so like ours as to be hard to distinguish, but their brains were little larger than those of apes, and if we could see them alive their physiognomy, while distinctive, would probably strike us as more apelike than manlike.

By a time probably not later than 500,000 years ago and perhaps earlier, gradual evolution from australopithecines had reached a stage that was human in a more restricted sense, belonging not only to the human family, Hominidae, but also to the same genus as ourselves, *Homo*. Doting and ambitious discoverers have given many different names to such early fossil men, including *Pithecanthropus* and *Sinanthropus*, but most of them are now usually placed in a single species, *Homo erectus*. Bodily anatomy and even physiognomy were now almost fully human, but to our eyes there was still a coarse or brutish cast of countenance because of heavy brow ridges over the eyes and a low, small brain case. The brain size was nearly intermediate between australopithecines (or modern apes) and modern man.

Finally, and still gradually, our own species, *Homo sapiens*, emerged. Although not entirely certain, it is now the usual opinion that the quite varied fossils known collectively as Neanderthal men belonged to *Homo sapiens* and only represent ancient races that were at first primitive (not so far removed from *Homo erectus*) and later somewhat aberrant. The more aberrant late Neanderthals became extinct as such, although it is probable that some of their genes survive.

So much for more or less direct knowledge of man's physical, anatomical origin. The main points are these:

1) Man evolved from apes also ancestral to chimpanzees and gorillas, but less specialized than the latter.

2) The divergence of man's ancestry from the apes was early marked by bipedalism and upright posture, with extensive correlations and implications in anatomy, habits, and capabilities.

3) Also early was divergent dental evolution, again with other implications, for example, as to diet and means of defense. It is not known whether posture and dentition diverged from the apes simultaneously or in which order.

4) Only after evolution of human posture and dentition was essentially complete did man's brain begin to enlarge beyond that of the apes. (Intelligence depends not only on size of the brain but also on its internal anatomy, and we do not know the internal anatomy of our fossil ancestors' brains. However, it is fairly certain that a species with average brain size as in apes could not be as intelligent as *Homo sapiens.*)

Now let us briefly consider the taxonomic, biological systematic nature of mankind as it exists today. First and most important is the fact that mankind *is* a kind, a definite and single species. A biological species is an evolutionary unit composed of continuing populations that regularly interchange genes by interbreeding and that do not or cannot have such regular interchange with other species. The definition clearly applies to mankind: all human populations can and, as opportunity occurs, do interbreed, producing fertile offspring and thus continuing the species and keeping it bound together as a unit. It is unlikely that, for example, a Greenland Eskimo has ever interbred with a South African Bushman, but since all intervening populations can and do interbreed they are nevertheless members of the same species. That species, *Homo sapiens,* is not connected with any other species by interbreeding.

Comparison of Eskimo and Bushman brings up the obvious (although occasionally denied) fact that the human species includes quite diverse races. A race is simply a population (or group of populations) that is genetically distinguished from others. The distinction is not absolute. It is unlikely that Negroes, for example, have any genes that do not occur in some white populations, or that whites have any genes absent in all Negro populations. The usual situation is that a race has certain genes and gene combinations that are more frequent in it than elsewhere, and therefore typical in that sense, but not confined to the race. Races always grade into each other without definite boundaries. Any two human populations, no matter how small or how large, differ in some respects, so that there is no fixed number of races. One could count thousands or two, and no matter how many are counted, there will be some populations and many individuals that do not clearly fit into one or another. Moreover, races are evanescent in the course of evolution. A given race may change, disappear by fusion with others, or die out altogether while the species as a whole simply continues its evolutionary course.

Races of man have, or perhaps one should say "had," exactly the same biological significance as the subspecies of other species of mammals. Widespread animals have local populations that live under diverse conditions and that may become temporarily and in part isolated from each other. They may then more or less accidentally have different proportions of genes (in stricter technical language, of alleles) from other such populations, and if the situation continues long enough, they will almost inevitably evolve somewhat different adaptations to local conditions. Primitive men were relatively few in number and relatively immobile, but they spread over enormous areas—the whole land area of the earth except for Antarctica and a few small islands. They evolved into races or, in better biological terms, into subspecies exactly as any other animal would have under those circumstances. Racial differentiation in man was originally geographic and, for the most part, adaptive.

That was the original biological significance of race. One must say that Negroes were biologically superior to whites, if reference is to prehistoric times, when the races were originating, and to African conditions, to which Negroes were biologically adapted and whites were not. At the present time race has virtually no strictly biological significance because of two crucial changes. First, human adaptation to different environments is now mostly cultural and is directly biological only in lesser part, so that the prehistoric biological adaptations have lost much of their importance. Second, tremendous increases in population size, in mobility, and in environmental changes brought about by man himself have the result that extremely few men are now living under the conditions to which their ancestors were racially adapted.

Evolution does not necessarily proceed at the same rate in different populations, so that among many groups of animals it is possible to find some species that have evolved more slowly, hence are now more primitive, as regards some particular trait or even over-all. It is natural to ask—as many have asked—whether among human races there may not similarly be some that are more primitive in one way or another or in general. It is indeed possible to find single characteristics that are probably more advanced or more primitive in one race than in another. For example, the full lips and kinky hair of some Negroes are almost certainly progressive traits in comparison with the more primitive, decidedly apelike thin lips and straight hair of most whites. However, that does not mean that whites in general are more primitive than Negroes or otherwise inferior to them. Over-all primitiveness and progressiveness in comparisons of different groups of animals are virtually confined to cases in which the groups are of different species, so that genes of the more rapidly evolving species cannot be transferred to the lagging species. Human races all belong to the same species and have generally had enough interbreeding so that genetic progress, as distinct from local adaptation, could and evidently did spread through the entire species. Only if some race entirely ceased to interbreed with any other would it be likely for it to fall behind and become definitely inferior. Let us hope that will not happen.

Regardless of the diversity of races, it is obvious that all men resemble one another much more than any of them differ from each other. They all share the basic qualities, anatomical, physiological and psychological, that make us human, *Homo sapiens*, and no other species that is or ever was. Something has already been said of anatomical peculiarities of *Homo sapiens* with respect to living apes and human ancestors. Here are some of the most striking human anatomical traits:

Normal posture is upright.

Legs are longer than arms.

Toes are short, the first toe frequently longest and not divergent.

The vertebral column has an *S* curve.

The hands are prehensile, with a large and strongly opposable thumb.

Most of the body is bare or has only short, sparse, inconspicuous hair.

The joint for the neck is in the middle of the base of the skull.

The brain is uniquely large in proportion to the body and has a particularly large and complex cerebrum.

The face is short, almost vertical under the front of the brain.

The jaws are short, with a rounded dental arch.

The canine teeth are usually no larger than the premolars, and there are normally no gaps in front of or behind the canines.

The first lower premolar is like the second, and the structure of the teeth in general is somewhat distinctive.

Given those characteristics, a museum curator could readily identify any specimen of *Homo sapiens* that was added to the collections, or that happened to walk into his office. However, we who are pondering the question "What is man?" must feel that these anatomical features, fully diagnostic as they are, yet do not amount to an answer adequate for our purposes. Even if we were defining, say, a species of mouse, the anatomical definition would not take us far toward understanding "What is mouse?" or, better, "What is mouseness?" unless we related the bodily mouse to the behaving mouse and the thinking mouse. Even thus, human anatomy reflects truly essential man-ness or human nature only to the extent that it is related to human activities and psychology. Already in *The Descent of Man* Darwin discussed such traits in which man appears to be most distinctive. His points, here greatly abbreviated and paraphrased, were as follows:

In proportion with his higher intelligence, man's behavior is more flexible, less reflex or instinctive.

Man shares such complex factors as curiosity, imitation, attention, memory, and imagination with other relatively advanced animals, but has them in higher degree and applies them in more intricate ways.

More, at least, than other animals, man reasons and improves the adaptive nature of his behavior in rational ways.

Man regularly both uses and makes tools in great variety.

Man is self-conscious; he reflects on his past, future, life, death, and so forth.

Man makes mental abstractions and develops a related symbolism; the most essential and complexly developed outcome of these capacities is language.

Some men have a sense of beauty.

Most men have a religious sense, taking that term broadly to include awe, superstition, belief in the animistic, supernatural, or spiritual.

Normal men have a moral sense; in later terms, man ethicizes.

Man is a cultural and social animal and has developed cultures and societies unique in kind and in complexity.

The last point, which some students now consider the most important of all, was least emphasized by Darwin, who was here mainly concerned with the relationship of social evolution to the origin of the moral sense. Darwin's general purpose was not to characterize *Homo sapiens* as the unique species that he is. The purpose was to show that the characteristics that make him unique are nevertheless foreshadowed in other animals, and that the evolution of man from other, earlier, quite distinct species is therefore plausible. We are no longer concerned with *whether* man evolved, because we know that he did. We are still very much concerned with *how* he evolved, with what is most characteristically human about him and how those characteristics arose. The list of traits discussed by Darwin is still valid from this somewhat different point of view.

That list should not be taken as involving so many separate and distinct things. These are aspects of the behavior, capacities, and accomplishments of a species that is characterized by all of them together and not by each or any one separately. They interact and interlock not only with each other but also with the previously mentioned physical or anatomical characteristics of man. For example, complex human societies, especially the modern industrial civilization rapidly spreading to the whole world, require specialization of activities by different members of society further involving manipulation of complex machines. Such specialization, which is nongenetic, requires individual flexibility and could not occur in a mainly instinctive animal. The machines are tools and could only have been devised by a reasoning, toolmaking animal. Invention also required manual deftness, which was provided by (and which also gave selective value to) the structure of the human hand, which required upright posture and could not have been acquired by a quadruped. Further evolution of the early cultural adaptations that led eventually to modern industry also had increased intelligence as a necessary concomitant, and that eventually required larger brains, which in turn involved change in skull structure and in stance—and so on. Even the changing pattern of the teeth can be related to this unitary complex.

Because all the specifically human traits are integrated within the whole that is human, and because each of the traits as well as their integration must have arisen gradually, it is somewhat questionable to speak of definite milestones or even of particular critical phases in the evolution of man. Yet there are three among these slow and co-ordinated changes that seem particularly basic for the concept of humanness. The most crucial single anatomical point is acquisition of upright posture and strictly bipedal locomotion. Most of the other main peculiarities of human anatomy either follow from that or are coadapted with it. The other two major factors are cultural, but are no less biological since both represent attainment and maintenance of biological adaptation by cultural means. They are toolmaking and language.

Extremely crude but unmistakable stone tools are found in the oldest rock strata containing indisputable members of the human family, nearly, if not quite, 2,000,000 years old. It will be difficult to authenticate still older and more primitive stone tools, because they must have consisted of natural pebbles or rock fragments picked up and used with little or no modification. It has long been maintained that deliberate manufacture of a tool is the distinctive human trait, since many other animals, even including some insects, use natural objects as tools but do not make tools. Now it has been found that chimpanzees may trim and shorten twigs or straws for use as tools, and although that simple behavior is almost too primitive to be called toolmaking, it sufficiently demonstrates that the capacity for toolmaking is biologically ancient and prehuman. If one wants a more diagnostic statement, it probably is true that man is the only living animal that uses tools to make tools. However, that trait would follow soon and inevitably once toolmaking really got under way. A stone used to knock flakes off an incipient stone ax is already a machine tool.

Ancient tools more perishable than stone are rarely preserved. Nevertheless, the course of increasing diversity and complication of tools can be followed well enough to demonstrate the gradual and inconstant but generally continual progress through prehistory. The tremendously accelerated progress in historic times is very well documented and is familiar to all of us in general outline, at least. The whole sweep from stone axes to electronic computers is a natural and comprehensible extension of the biological capacities of an unusual species. It is uniquely wonderful, and yet, lest we stand too much in awe of our own products, let us remember that a digital computer is merely a rapid and automated tool for what amounts to counting on fingers.

As posture is focal for consideration of man's anatomical nature and tools are for consideration of his material culture, so is language focal for his mental nature and his nonmaterial culture. Language is also the most diagnostic single trait of man: all normal men have language; no other now-living organisms do. That real, incomparably important, and absolute distinction has been blurred by imprecise use of the word "language" not only in popular speech but also by some scientists who should know better, speaking, for example, of the "language of the bees."

In any animal societies, and indeed in still simpler forms of aggregation among animals, there must be some kind of communication in the very broadest sense. One animal must receive some kind of information about another animal. That information may be conveyed by specific signals, which may be of extremely diverse kinds both as to form and as to modality, that is, the sensory mode by which it is received. The odor of an ant, the movements of a bee, the color pattern of a bird, the howl of a wolf, and many thousands of others are all signals that convey information to other animals and that, in these and many other examples, are essential adaptations for behavioral integration in the species involved.

Human language is also a system of interpersonal communication and a behavioral adaptation essential for the human form of socialization. Yet human language is absolutely distinct from any system of communication in other animals. That is made most clear by comparison with other animal utterances, which most nearly resemble human speech and are most often called "speech." Nonhuman vocables are, in effect, interjections. They reflect the individual's physical or, more frequently, emotional state. They do not, as true language does, name, discuss, abstract, or symbolize. They are what the psychologists call affective; such purely affective so-called languages are systems of emotional signals and not discourse.

The nonlanguage of animal interjection is still present in man. In us it is in effect not a part of language, but the negative of language, something we use in place of speech. In part we even use the same signals as do the apes, a fact already explored to some depth by Darwin in another of his basic works, *The Expression of the Emotions in Man and Animals.* Much more is now known about such expressions in animals, and particularly in our closer relatives the apes and monkeys, and it is not surprising to find that the nonlinguistic, affective system is particularly complicated in them and has not progressed but may even have retrogressed in man. Still we do retain that older system along with our wholly new and wholly distinct system of true language. It is amusing that the human affective interjectional reaction to a bad smell is practically the same as in all other primates, down even to the most primitive.

Darwin's study and many later studies sought to trace the evolutionary origin of language from a prehuman source. They have not been successful. As Lancaster has said, "The more that is known about it [that is, communication in monkeys and apes], the less these systems seem to help in the understanding of human language."

Many other attempts have been made to determine the evolutionary origin of language, and all have failed. Because language is so important for any concept of man and because this is an interesting example of methodology and limitations, it is worthwhile to consider some of these futile attempts. One, fairly obvious once the idea of linguistic evolution had arisen, was by comparison of living languages. One result was a supposed genetic sequence: (1) isolating languages, like Chinese, which string together invariable word roots; (2) agglutinating languages, like Mongolian, which modify roots by tacking on prefixes and suffixes; and (3) flexional languages, like Latin, which modify by (partly) internal changes in words. The trouble is that these categories are not really distinct and, especially, that they did not historically occur in this sequence. For example, Chinese was probably flexional at one time and is now becoming agglutinating with a possibility of becoming flexional again. English was flexional until quite recently and is now mostly isolating with a strong dash of agglutination. Moreover at the present time no languages are primitive in the sense of being significantly close to the origin of language. Even the peoples with least complex cultures have highly sophisticated languages, with complex grammar and large vocabularies, capable of naming and discussing anything that occurs in the sphere occupied by their speakers. Tales of tribal natives who cannot count beyond four and who have vocabularies of only two or three hundred words betray the shortcomings of gullible travelers, not of the natives.

Another approach is to follow back directly historical records, which cover several thousand years for some European, Asiatic, and North African languages. It is then possible to project still further and to reconstruct, for example, a proto-Indo-European anterior to Sanskrit. But this still leaves us tens or hundreds of thousands of years—perhaps even more—from the origin of language. The oldest language that can reasonably be reconstructed is already modern, sophisticated, complete from an evolutionary point of view.

Still another attempt, which now seems very naïve, is through the ontogeny of language, that is, the acquisition of language by children. That approach, too, is invalid, as is further explained in Chapter Eight.

It is a tempting hypothesis that the time, at least, of the origin of language might be determined by structural characteristics in fossils. One rather elaborate attempt departed from the fact that all linguistic phonetic systems, varied as they are, depend in part on the shape of the lower jaw and the hard palate, anatomically quite different in typical members of the human and the ape families. It was postulated that speech began when these anatomical parts reached human form, which was in the australopithecines or somewhat earlier. But the postulate is clearly wrong. Audible signals capable of expressing language do not require any particular phonetic apparatus, but only the ability to produce sound, any sound at all. Almost all mammals and a great number of other animals can do that. Moreover, a number of animals, not only birds but also some mammals, can produce sounds recognizably similar to those of human language, and yet their jaws and palates are radically nonhuman. A parrot is capable of articulating a human word but is completely incapable of understanding what the word means.

Given any method of sound production, the capacity for language depends not on characteristics of the sound apparatus but on the central nervous system. Speech is particularly connected with the left temporal lobe of the human brain, as shown, for example, by the fact that ability to speak is generally lost if that lobe is severely damaged. The gross development of the lobe can be seen in plaster casts of the insides of fossil skulls, and that, too, has been proposed as a means of determining whether or not a given fossil individual could speak. But all mammals have left temporal lobes, some smaller and some larger. Those with smaller lobes do not speak just a little and those with larger lobes more. There is no graded sequence: normal men speak completely; other animals, whatever the relative size of their temporal lobes, do not speak at all.

The essential anatomical and physiological basis of speech is nevertheless in the structure and function of the brain. That basis is not fully known, but it evidently involves not just a language center, such as might be localized in the temporal lobe, but an intricate and widespread system of associative connections throughout much of the brain. (The nature or presence of these connections cannot be determined in fossils.) Thus sensations of any kind derived from an external object or event can be generalized according to similarities with others. Each kind can then be associated with a distinctive symbol, which does not resemble the object or event at all but which arbitrarily stands for it. That symbol, a supreme element in the nature of man, is the word, and it is not surprising that words meaning "word," abstraction and symbolization on still another level, have acquired such mystical and philosophical overtones. (Λόγος!)

It is still possible but it is unlikely that we will ever know just when and how our ancestors began to speak. Yet it is certain that this ability depends on physical, structural, and chemical characteristics of the nervous system which evolved from our nonspeaking ancestors under the force of natural selection. The capacity for this unique kind of symbolization is quite general. It does not determine what symbol will be used for a given concept, but that any symbol can be associated with any concept. Thus we are all using exactly the same genetic capacity and symbolizing the same concept when various of us say "woman," "Weib," "femme," "mujer," "zhenshchina," or "imra," depending on whether we happen to have been raised in England, Germany, France, Spain, Russia, or Egypt. The words do not resemble each other and even less resemble the concept they stand for. Moreover, they can be written in different ways, as in Latin, Arabic, or Chinese characters, that do not resemble each other and that have no physical resemblance to the spoken words. They can even be associated with some symbol that is not verbal at all, as in this example with the simplified representation of Venus's mirror that biologists use to designate females: ♀.

Language has become far more than a means of communication in man. It is also one of the principal (although far from the only) means of thought, memory, introspection, problem-solving, and all other mental activities. The uniqueness and generality of human symbolization have given our mental activities not only a scope but also a quality far outside the range of other animals. It keeps us aware, to greater extent than can otherwise be, of past and future, of the continuity of existence and its extension beyond what is immediately sensed. Along with other peculiarly human capacities, it is involved in what I consider the most important human characteristic from an ethical point of view: foresight. It is the capacity to predict the outcome of our own actions that makes us responsible for them and that therefore makes ethical judgment of them both possible and necessary.

Above the individual level, language and related powers of symbolization make possible the acquisition, sharing, and preserving of knowledge far beyond what would be possible for any single individual. That is an indispensable element in all forms of human social organization and cultural accomplishment, even the most primitive.

It is obvious that I have by no means touched on all aspects of the biological nature of man. That would be impossible in one essay by one author. Those familiar with recent developments in biology may particularly miss reference to molecular biology and especially to the compound called DNA, now known to be largely involved in heredity and also in control of biochemical activities in cells. Those subjects are extremely fascinating at present and may be portentous for the future. However, in my opinion nothing that has so far been learned about DNA has helped significantly to understand the nature of man or of any other whole organism. It certainly is necessary for such understanding to examine what is inherited, how it is expressed in the developing individual, how it evolves in populations, and so on. Up to now the triumphs of DNA research have had virtually no effect on our understanding of those subjects. In due course molecular biology will undoubtedly become more firmly connected with the biology of whole organisms and with evolution, and then it will become of greater concern for those more interested in the nature of man than in the nature of molecules.

Finally, it should be pointed out that although man is a unique animal and although we properly consider his nature in the light of his peculiarities, he also has many nonpeculiarities. Man is not *merely* an animal, that is, his essence is not simply in his shared animality. Nevertheless he *is* an animal and the nature of man includes and has arisen from the nature of all animals. Indeed if all the material characteristics of man could be enumerated, it would surely be found that the vast majority of them also occur in other animals. In fact, at the level of molecular structure and interaction, information storage and transfer, energy transactions, and other defining characteristics of life, man is hardly significantly different from a bacterium—another illustration of the fact that that level of study is not particularly useful in considering the nature of man.

Like other animals, man develops, is born, grows, reproduces, and dies. Like other animals, he eats, digests, eliminates, respires, locomotes. He bends the qualities of nature to his own ends, but he is as fully subject to nature's laws as is any other animal and is no more capable of changing them. He lives in biological communities and has a niche and an ecology, just as do robins and earthworms. Let us not forget those aspects of man's nature. But let us also remember that man stands upright, builds and makes as never was built or wrought before, speaks and may speak truth or a lie, worships and may worship honestly or falsely, looks to the stars and into the mud, remembers his past and predicts his future, and writes (perhaps at too great length) about his own nature.

CHAPTER SEVEN

Biological Aspects of Race

"That Charlie Bird, he's half Cherokee and half white, and them's two bad breeds."—Ascribed to BILL DOOLIN by E. M. RHODES

Some anthropologists have maintained that human races do not exist. UNESCO has issued a statement that race is a strictly biological concept without psychological or cultural connotations. If the former were true, the title "Biological Aspects of Race" would be meaningless; if the latter, it would be redundant. In both cases, the spokesmen had their hearts in the right place but their eyes closed. It is a fact of biology that human races exist; it is a fact of life that their existence involves not only biology but also psychology, sociology, economics, politics, theology, and plain old-fashioned hysterics. The implications of race in all those and other fields depend on conceptions, more often misconceptions, as to the biological aspects of race. Rational approach to the problems and the conflicts involving race requires better knowledge of those aspects.

In the first place, mankind is a species with the technical name *Homo sapiens,* just as pumas are a species, *Felis concolor,* and ravens are a species, *Corvus corax.* A relevant definition of a species is that it is a population (or complex of populations) of similar organisms maintaining continuity and unity through the generations by interbreeding among themselves and maintaining identity by relative or, more often, absolute absence of interbreeding with other such groups (species). That definition, or any other acceptable in modern biology, applies perfectly to the human species. Men, all men, are brothers in the literal sense of belonging together in the basic biological unit, a species.

All species vary; in fact no two individuals of any species are ever *exactly* alike. If you think that all pumas or all ravens are alike, you have not looked at them closely. Obviously *Homo sapiens* is no exception to that rule. Much of the variation is due to developmental interactions with the environment, always within limits set by heredity. Such variation is also of biological importance, but for the present topic only variations in total heredity, that is, in the genotype, are fully relevant. The genetic structure of specific populations and the processes of sexual reproduction make it virtually impossible for any two bisexual organisms to have precisely the same genotype. Among humans, identical twins are the only exceptions.

When hereditary characteristics within one population have two or several fairly distinct states, they are called "morphs" (from the Greek for "forms") and a population with known morphs is called "polymorphic" ("many-formed"). For example, all studied local populations of humans are polymorphic for blood groups. There are many blood groups, but even in the A, B, AB, and O groupings more than one type occurs in any population. Although simple observation may be deceiving and although the morphs intergrade, most populations in Europe or of that origin are polymorphic for hair and eye color. Polymorphism is a normal characteristic of biological populations, and it is not at all special to man. It occurs in innumerable species throughout both animal and plant kingdoms—certainly in most and probably in all of them if we knew all their characteristics. As just one other example, at least thirteen quite distinct color patterns occur in males of wild populations of one species of platyfishes.

Groups of individuals with the same morphs have often been called "varieties." The point here is that a variety, in this sense, is not a biological population, unit of classification, or evolutionary entity. It is only one kind of variant *within* such a population, unit, or entity. Its status is no different from that of other genetic variations within populations. It has nothing directly to do with race, and confusion of the two—variety and race—has been a source of serious misunderstanding. This is one of the few points on which even Darwin was confused.

If a species is examined over its full geographic extent, another kind of variation is evident: the populations in different places are, on an average, different in their hereditary characters. Exceptions may occur among species consisting of one small, local, interbreeding group, but only a few and doubtful exceptions have been found among adequately studied species with geographically dispersed breeding groups. Again man is obviously no exception. In spite of mobility and mixture, the average character of populations in, say, Peking, Calcutta, Oslo, Accra, and Taos Pueblo is obviously different. This is precisely the same biological phenomenon, in origin, as the existence of average differences between, for example, populations of brown bears in the Pyrenees and in Alaska or of song sparrows in New York and Oregon.

These distinguishable breeding groups within a species, in *Homo sapiens* or in any other species, are races. A species that includes clearly distinguishable races is called "polytypic," and *Homo sapiens* is a fairly characteristic polytypic biological species or, at least, was so in more primitive times when the races now distinguished were coming into being. Polymorphism is quite different from polytypism: morphs occur within a local population and are not systematic units in biology; the races of polytypism are local populations and are systematic units. Both races and species are such units but these, too, differ in a fundamental way: species are highly or, as in *Homo sapiens*, absolutely isolated genetically, but adjacent races are not genetically isolated—they can and do interbreed regularly or intermittently with other races of the same species. Recognition that mankind is one species and that races are nonisolated or incompletely isolated populations within a species is the beginning of wisdom on racial problems.

For one thing—an important one—it is clear that the concept of a "pure" race is biological nonsense. It is hard to see what this word might be taken to mean in terms of modern genetics, and no assignable meaning applies to any real races, human or otherwise. A genetically homogeneous breeding population is impossible; such a thing certainly never has existed and never will. A race all of whose members have *any* genetic trait entirely absent in adjacent races is also almost out of the question, because adjacent races interchange genes. Races always intergrade. They are always "impure" if that loaded word is taken to mean genetic heterogeneity. They are never clear-cut, unequivocally definable entities, as species normally are. Races are populations defined by genetic relationships. They cannot be defined in terms of individuals or of personal characteristics.

It further follows that there is no given number of human races. Any two populations, such as the peasants of adjacent Alpine valleys, differ measurably if measurements and averages are taken to enough decimal places. They and the untold thousands of other such local groups could legitimately be called races if anyone cared to. In aboriginal America alone there were thousands of Indian tribes each of which could readily be defined as a race. On the other hand all of mankind could also be classed in just two races on legitimate, even if not particularly clear or useful, criteria. The numbers currently recognized by different anthropologists, or even by one anthropologist at different times, usually vary from three to about thirty. There is no "right" number; it all depends on the taste and purposes of the namer.

In biology the term "race" is not a systematic category but is used for any local infraspecific breeding group that is conveniently distinguished for purposes of a given study. The term "subspecies" is used in a technical sense for a race considered sufficiently distinct, uniform, and widespread to merit a Latin name. There are also some strong differences of opinion and judgment as to this usage. The so-called major human races of many anthropologists are usually taken as three to six in number, and they correspond well enough with biological subspecies. It would perhaps reduce some emotionality if these groups, such as the Caucasoids, Mongoloids, and Negroids, were called subspecies instead of races. However, there are many smaller groups that cannot be assigned in a really valid biological way to one of these major divisions, and it is equally good biology to apply the term race loosely to such groups of all sizes and degrees of distinction.

An enormous amount of study has been devoted to problems as to how and why animal races evolve and the significance of differences among them. The subject is exceedingly intricate and we are far from complete understanding, but much is known and that much is enlightening as regards human races. Animal races normally arise in different geographic areas as populations that become more or less distinct genetically. The distinction depends on no absolute genetic trait but on different incidence or frequency of genes. In a small population and in early stages of racial differentiation the differences may arise more or less by chance. Eventually, however, if the process continues, the differences established and maintained are usually adaptive. In a widespread species (few are as widespread as man) local populations live in different environments. By natural selection they tend to acquire traits—racial characteristics—adaptive to those environments.

There is no doubt that human races arose in just the same way. They certainly evolved as populations in different geographic regions, as is still obvious even though human mobility has blurred and complicated the picture. At a time of semi-isolation among small widely dispersed tribes, it is likely that chance fixation of some racial traits occurred. It seems certain, however, that many or most prominent racial characteristics were originally adaptive. Curiously enough, there is not much really firm knowledge on this extremely important point. One trouble is that incomplete and mistaken concepts of the biological nature of race long led the anthropologist to study the wrong racial traits, or to study them in the wrong way. Now it has been found that supposedly nonadaptive characters, such as blood groups, do in fact have adaptive or inadaptive aspects, such as immunity or susceptibility to certain diseases in the case of blood groups. Many more obvious traits now appear as also more obviously adaptive. Skin color correlates closely with the intensity and kind of natural radiation in various areas. Body build correlates closely with other climatic factors, especially temperature.

Part of the lesson here is the absurdity of thinking of such adaptive characteristics in terms of "better" or "worse," "higher" or "lower." It makes sense to think of Negroes as superior to all others if reference is to the time when major race differentiation was occurring and to the particular region where the Negroid race evolved. But Mongoloids were superior at the same time in a different region, and so on for all races. Each was adaptively superior in some way and degree where and when it arose. Now, for reasons to be specified later, even that relative kind of biological superiority has lost most of its meaning.

Such traits as light or dark skin color have biological importance only as they do, or rather did, represent physiological adaptation. It is a psychological phenomenon that such easily observed characteristics may act as releasers for hatred and hysteria. That is connected with their being taken as labels for less easily observed and more deeply important traits that are supposed to be correlated with them. Most essential of such traits are those of intelligence, talent, and personality. The rationale of racial labeling, insofar as it has one, is that the visible racial trait, notably skin color, is supposed to accompany and hence label intangible traits, which are good if the label is the same as ours and bad if it is not.

Not much is yet known about distribution of such intangibles in nonhuman animal races. Yet we do know that the labeling concept, as applied by racists, is biologically invalid. If, for example, one animal race has darker fur than another, it cannot be judged beforehand that it will also differ in any other character, such as having longer or shorter ears. By careful measurement of large samples of each we might find empirically that race A has on an average darker fur than race B and also has on an average longer ears. Even then we could not take these as individual labels; they are only populational averages or tendencies. Since color of fur and length of ears vary in both populations and vary independently of each other, it is quite expectable that some individual with dark fur will also have short ears. Association of a physical variable with a psychological trait cannot be assumed to have a firmer basis, and may not be found at all.

Now as regards human races, we have not even taken the first step of establishing empirically an average association of label and contents, of physical trait and mental or personality traits. It is true that various psychological tests have been applied to different racial groups, although not yet very thoroughly and systematically. It is also true that these tests have reliably indicated average differences among some racial groups in whatever it is that some particular intelligence or personality test is measuring (often debatable). But the statistically significant differences are almost all culture-dependent to considerable and uncertain degree. So far as anyone can be sure, what they have revealed are differences between, for example, the Chinese and African civilizations or ways of life and not between the Mongoloid and Negroid races. They do not establish any reliable association of biological traits. Tests that are most nearly culture-free generally indicate such wide overlaps in variation of compared groups that differences in averages are unreliable and differences between individuals virtually meaningless as regards association with race.

There are biological reasons why significant racial differences in intelligence, which have not been found, would not be expected. In a polytypic species races adapt to differing local conditions but the species as a whole evolves adaptations advantageous to all its races, and spreading among them all under the influence of natural selection and by means of interbreeding. When human races were evolving, it is certain that increase in mental ability was advantageous to *all* of them. It would, then, have tended over the generations to spread among all of them in approximately equal degrees. For any one race to lag definitely behind another in overall genetic adaptation, the two would have to be genetically isolated over a very large number of generations. They would, in fact, have to become distinct species; but human races are all interlocking parts of just one species.

Similar conditions apply to the concept that some races are biologically more primitive than others, that is, more like their remote ancestors. That can happen as between different related species, although even there it rarely if ever happens that one species is more primitive in *all* its characteristics. Among subspecies or races, differences in all-over primitiveness are generally negligible, and that is true of human races. Certain traits in the whole adaptive complex may be more primitive.

The greatest biological peculiarity of man is that he has developed a social culture different in kind from that of other species and enormously more complex. This has become his principal means of adaptation, and it has changed the role and significance of races in this unusual species. The differences among races are now largely irrelevant from a biological point of view. In a state of nature, Negroids are still better adapted to the Central African climate than Caucasoids, for example, but few now live in a state of nature. Under existing cultural adaptations, Negroids and Caucasoids can live and breed there about equally well. A further point is that the primitive environments, adaptation to which was the origin of the major human races, no longer exist. A few small groups, perhaps some remote Amazonian tribe, may still be living under conditions similar to those of their ancestral biological adaptations. That will not last long, and already the vast majority of men live in environments so different from their ancestry that their racial adaptations now have little biological significance.

The major racial differentiation dating from the childhood of *Homo sapiens* is thus relictual, more a memento of the past than a functional aspect of the present. That is evident in another way and has other present significance. Races of animal species arise in different areas, as did those of man, and if they are found in the same area this is temporary. Two closely similar groups are not ecologically compatible. If associated in the same area and ecological situation either they merge or one moves out or dies out. Yet in human populations there are now almost no areas occupied exclusively by the descendants of only one of the major prehistoric races. That is simply evidence that those groups, although to a somewhat limited extent still physically recognizable, are not now functioning biologically as races.

In some places (Hawaii and Mexico, for example) descendants of different prehistoric races are merging and producing what are, in effect, new functional geographic races. In others (Mississippi, for example) there has also been extensive interbreeding but the prehistoric races are still kept somewhat distinguishable by social, not otherwise biological, processes. Undoubtedly other new races will arise from diverse hybridization of old races—that is a biologically usual process and was also involved in the origin of the prehistoric races now relictual. That this will produce a uniform, not at all polytypic species is highly improbable, hardly possible within the next several thousand years. Because of human social stratification and prejudice, it is also possible that new races will arise not on the geographic basis usual in other species and originally active in this one, but by some form of social segregation of breeding groups. That, too, is unlikely to become significant within the next several thousand years. Hindu castes existed as such breeding groups for at least 3,500 years until recently abolished, but their racial differentiation was barely incipient.

A final point is that the biological situation most apt for evolutionary progress is a species with considerable variation (polymorphism, broadly speaking) within populations and with local differentiation (polytypism) of populations (races). That also holds analogously for cultural processes, which are now the main channel of human advance: diversity both within and between cultures is a condition for progress.

CHAPTER EIGHT

Language in Evolutionary Perspective

> "The vast and variegated realms of language and psychology cannot be covered in a few general remarks, however wise they may be, and no one—with the possible exception of a physicist I know who likes to lecture on The Universe and Other Things—no one could seriously hope even to state, much less to solve, all the problems that they pose."
> —G. A. MILLER

If you ask a scientist, "What is language?" his reply is likely to reveal much about his science and his attitude toward it. Brass-instrument or behaviorist psychologists, who have in common at least the point that they consider their subjects more as experimental objects than as people, may reply that language is "one example of the many devices used by living organisms to influence the behavior of other living organisms"—I am quoting one of them verbatim. Ethologists and other students of comparative behavior along with many linguists, especially those with an analytical or engineering slant, usually think of language as a system of communication. That is also usually true of social psychologists and sociologists, but there is a difference in how they use the concept of "communication." The ethologist may use the word to mean the passage, even involuntary, or nonfunctional, of any kind of information to a recipient. Social communication is a form of intercourse, a functional (or useful) and often voluntary interchange of information.

Still others consider language as the necessary medium of thought, and not only the medium but also the molder, restrainer, or distorter of thought. That was the concept of the late B. L. Whorf, most eminent of amateur linguists, and it has long been supported and twisted into protean forms by a school of philosophy. I think it contains a basic fallacy and regret that its discussion would take me too far afield here. I shall only note that some extremely verbal people have been astonished to learn that it is possible to think at all without words! However, language is *a* means of thought, if far from being the only one or placing limits on what can be thought.

Poets and novelists have other ideas on the function of language, and so do politicians, although they usually agree with the behaviorists—which, in fact, they are, on a technological level. Some shamans and priests have their concepts, too, not only (like the theoretical and practical behaviorists) that language can influence or force action by other humans, but also that it can influence or force action by the gods. In all, concepts of language are so varied that the question could be made the basis for one of those not-so-innocent parlor games that aim to lay bare the personality of a guest.

It happens that I am an evolutionary biologist and systematist, and I, too, have my conditioned answers. From a systematic point of view, language is the most entirely unique, the most completely diagnostic characteristic of *Homo sapiens*. From an evolutionary point of view, language is a biological adaptation essential in the way of life specific to that species. From both points of view, we would like to know just what it means and how and when it happened that our ancestors acquired language and that no other living animals have it. Throughout discussion of those topics, it is to be borne in mind that language does not have one or the other of the functions earlier mentioned: it has them all at once, and others, too.

The statement that language is absolutely confined to man arouses dissent and requires definition. Because language is, among other things, a means of conveying information, it has been assumed that any means of conveying information is language within the scope of information theory—a subject so "in" that use of its jargon is a status symbol for many of the more hep scientists. But information, as the term at its broadest is used in information theory, is received and in some way processed and utilized by all organisms without exception. At least, something in the physical environment yields "information" to the organism—be it a bacterium, an orchid, or a man—which reacts in some way. This is no real problem, for few of the theory's most rabid devotees would call that "communication." For communication one must have, if not literally then analogically, the sequence laid down in Claude Shannon's original, mechanical form of communications theory: source-encoder-channel-decoder-receiver.

Communication must at very least involve transfer of information from one organism as source to another as receiver, but even that leaves us in doubt as to the proper boundary of the field of communication and very far from that of language. If a fox and a rabbit see each other and one pursues where the other flees, information has been exchanged within the acceptance of the term, but that is dubiously communication and certainly is not language. Let us move up another notch, a large one, to the justly much-publicized so-called "language" of the bees. Exploring or foraging honey bees, on returning to the hive, make movements and buzzes that convey fairly exact information about the directions and distances to sources of honey and pollen. That is certainly communication, useful on both sides, biologically adaptive for the species, and clearly involving encoding and decoding nearly analogous to Shannon's mechanical system. Indeed the encoding is so elaborate and, to humans, so unexpected that its interpretation (the first essentials by Karl von Frisch) required considerably more ingenuity, but incomparably less expense and infinitely less peril to mankind, than splitting the atom.

Communication, yes. Language, no. A system of communication, even though symbolic and coded, cannot meaningfully be called "language" if it is largely or wholly instinctive, altogether inflexible, essentially involuntary, and capable of conveying only a modicum of information on one single topic. The "language" of the bees and similar phenomena among other animals are so unlike true language and so completely different in origin that we can learn nothing about the nature and origin of human language from them.

We seem to come nearer to language and to clues to its origin when we note that many animals, especially those more closely related to man, make a variety of sounds. Not in all but in many instances those sounds vary in a systematic way associated with particular circumstances of the animal uttering them. They contain information about those circumstances and the animal's reaction to them, and they may be used, often involuntarily but sometimes purposefully, to communicate that information to another animal. Although the psyche of nonhuman species may be considered impenetrable, every owner of a pet knows that dogs and cats emit different sounds in different situations and that they frequently use those sounds to communicate emotions or desires to other members of their own species and to members, such as their owners, of other species. The same kind of communication has been found in many wild animals, especially mammals, possibly with as many as fifteen or twenty different signals—the number is necessarily indefinite because, apart from problems of observation and analysis, the signals are not discrete units analogous to words.

Here there is an evident resemblance to language. A good many evolutionary biologists—including Darwin—and linguists—including Jespersen, one of the great pioneers—have considered the resemblance to be an identity, at different levels of evolutionary progress, and the animal signals to be the origin of human words. Lately, however, it has become increasingly evident that this is not so, and ideas of the occurrence of true, albeit exceedingly primitive, language in living nonhumans have been rather reluctantly abandoned. As one of the present-day students (Jane Lancaster) of communication in monkeys and apes has remarked, "The more that is known about it, the less these systems seem to help in the understanding of human language."

What is the difference? For one thing, the great majority of (nonhuman, to be understood) animal utterances are affective or interjectional: they are evoked by the physiological and emotional state of the animal at the moment. They do not name, describe, or generalize. They are not symbolic. They do not correlate with anything not actually, physically present at the moment of utterance. They are nonproductive: each vocalization is an isolated unit and all correlate specifically with a very small number of particular situations; they cannot be extended or combined to convey information on more numerous or more complex situations. They do not permit of increased learning and cannot express a body of knowledge, hence do not transmit a way of life or the lore of a culture.

All of those things that animal vocalizations do not and cannot do, human language does with extraordinary perfection. Every language names, describes, and generalizes everything its speakers ever have occasion to communicate or, indeed, to think about. It does so by symbolizing all things, actions, and abstractions. It can and does refer to things not present, to the past, to the future, indeed to things (or concepts) that are not thought of as ever occurring in reality. It is, literally, infinitely productive: there are no limits to its capacity for extension and combination to apply to novel situations, to things never thought or communicated before. It enormously facilitates and broadens the possibilities of learning. It is the principal form of social inheritance and a main element in the maintenance, functioning, modification, and transmittal of cultural characteristics.

Some linguists who accepted the (indeed almost obvious) validity of those contrasts nevertheless used to insist that all that was needed to transform animal interjections into human language was transfer of function from the affective to the objective. That would not suffice, since it has been shown that there is vastly more to language than its objectivity, but in any case the argument was fallacious. As Ernst Cassirer pointed out long since, the transfer so lightly assumed is exactly the problem. There is no tendency for it to occur in living animals and no evidence or established likelihood that it ever occurred. Moreover, immense efforts have been made to effect such a transfer in comparatively highly intelligent experimental animals, especially dogs, apes, and recently dolphins. Some remarkable claims have been made, but all are clearly untrustworthy or misinterpreted. In reality all these experiments tend to confirm the absolute, qualitative distinction between general animal communication and specifically human language. Incidentally, Aristotle made this distinction; he was not always wrong.

The capping evidence is that both kinds of utterances are still used by man, and that they remain distinct in character and function. The difference is that between saying "Ouch!" and "Fire is hot"—an interjection expressing the affect of the moment, and a symbolically coded, representational communication of a timeless generalization. Indeed interjections of that kind, which is the only kind known in other animals, are not language but antilanguage. They occur in lieu of, not as parts of, speech. Because we have this affective system in common, we can indeed learn something about human communication from comparisons with our nonhuman relatives, but not about language.

It should now be clear why the biologist regards language as an evolutionary adaptation, species-specific to *Homo sapiens*, and indispensable to the whole way of life of that species. The particular and unique way that man has for coping with his fellows and his environment would be absolutely impossible without language. As regards this point, we need not enter the dispute whether language led to human cultural adaptation or whether the beginnings of that adaptation led to the acquisition of language. It is improbable that either of those statements is true. Species-specific human culture and language, as part of that culture, must almost of necessity have arisen inseparably together.

It has already been mentioned that thought can occur in man—and surely in many other animals, too—without language. Nevertheless, all normal humans—but no other animals—do use language as a means of thought, some almost exclusively, others as a necessary auxiliary. Still more important, a fully human system of communication requires language. Even if, as often happens to me, my thought is wordless, I describe it in words or translate it into words when I pass it on, as I am doing at this moment. An individual object, no matter how complicated, can be indicated or described by drawing a picture, but pictures can be used to indicate actions, to generalize, or to abstract only as they become symbols and not portraits. Then the pictures themselves become a form of language and transcend mere pictorial representation. That is the origin of all alphabets, in which the pictures finally become symbols for sounds, which were symbols for other things. Pictures can alternatively become a symbolic language in themselves with no reference to sound. That is true of Chinese ideographs, which can be understood equally well and used for full communication by readers who have no spoken word in common, but which must therefore be learned by both as a language additional to speech.

One of the enticing side tracks to which I can point here but which I cannot follow far is a relationship between language and ethics. The point is debatable, but I have maintained at length elsewhere and I only repeat here that nothing of ethics is inherent in the evolutionary process or the history of life before the appearance of man. It simply makes no sense to speak of moral good or bad or of ethical responsibility except in reference to the ethical animal, *Homo sapiens.* As C. H. Waddington and others have shown, the talent and the need for ethicizing have arisen by biological evolution as a species-specific adaptation of this species. The same is true of language, and these two human phenomena are interdependent. Most essential is the fact that language took thought, learning, and communication out of the bounds of the present and made possible projection backward and forward in time. It facilitated, if it did not cause, foresight in man. It is foresight, the ability to predict the outcome of his actions, that makes man responsible for those actions and for chosing alternatives according to criteria of ethically good or bad probable outcome. Ethical concepts are also communicated mostly by language. It is improbable that ethics could or would exist without language.

From all those considerations it follows that language, in any of its possible forms, is largely definitive of the human condition, essential for human thought, for human communication, for human learning, and for the new form of inheritance and of evolution that has arisen in man. The old evolution by which man arose was strictly and directly biological. That continues, willy-nilly, but now it is accompanied by a new evolution, which is psycho-social, in

Julian Huxley's term, or cultural. Yet this evolution is an outcome of the old, biological evolution and is a continuation of it by other means.

How can language be considered an outcome of biological evolution, hence genetic in determination, and a form of biological adaptation, when in fact use of one language or another is a cultural trait learned anew by each individual? There is no real difficulty here. What has arisen by biological evolution and is genetically inherited and determined is the *capacity* to learn and use a language. That is already clear enough from the fact that no other species is inherently, genetically capable of learning a language. An anecdote by William Mann, late director of the national zoo, bears on this and also on the merely physical capacity for speech. Mann trained a myna bird to say, "How about the appropriation?" and then introduced the Director of the Bureau of the Budget to the bird. That was linguistic communication at a sophisticated level, but the bird remained naïve!

The linguistic ability that evolved and that is biologically determined is not a form or co-ordination of mouth, throat, tongue, and so on—the myna had all that—but a mental capacity, a cerebral structuring, which no nonhuman has. It is now known that there are several specific, localized areas in the human brain that are implicated in the use of language. Even that is not enough. There must also be a complex and unique set of neural connections so that perceptions in any modality (such as sight, sound, or other) can be linked in a fixed way to a symbol, which may also be in any modality. This extraordinary system, still very inadequately known in anatomical and physiological terms, can link *any* concept or perception to *any* symbol.

There has been a long search for knowledge as to just how language arose. The search has been largely abandoned now because it was not getting anywhere. It was not getting anywhere principally because it tried to discover a specific first language, or kind of language, which is probably impossible and is not very important, instead of the origin of the basic capacity for language, which may be possible and is certainly important.

Among the oldest theories was that of Plato, who felt that for a word to mean a thing there must be a real connection between the word and the thing. (Later theologians were to argue whether the names given by Adam in the Garden of Eden were the "real" names of the beasts!) Plato sought the connection in imitation or onomatopoeia. That hypothesis had defenders right down through the nineteenth century and may possibly have some today. It is nevertheless nonsense. What, for example, does a woman sound like (do not all answer at once!), and among the words "woman," "Weib," and so on (as above), which sounds most like a woman? How can all sound like women? Democritus, in this and some other respects not quite as silly as Plato, thought that human speech derived from animal cries. I have already noted that this hypothesis, in more sophisticated form, has lasted to our day, but that it is almost certainly wrong.

A more concrete and, as it seemed for a time, more promising approach arose with the development of comparative philology in the nineteenth century. Unequivocal records made it quite easy to trace, for example, French and Spanish to a common origin in vulgar Latin. With more labor and uncertainty, a postulated ancestor of classical Latin and Greek not far removed from ancient Sanskrit could be reconstructed, and still more remotely there is a largely hypothetical reconstructed proto-Indo-European, ancestral not only to all the languages mentioned but also to innumerable others from Celtic to Persian and onward. But there the process stops. Reconstructed proto-Indo-European has no definable connection with Altaic, Semitic, Bantu, or a host of other languages, and no reconstruction can conceivably be close to the primeval speech of *Homo*.

Let us then compare modern languages, arrange them in sequence from most to least primitive, and extrapolate on that basis. But what is primitive? My neighbors the Navahos (I am writing in Arizona) speak a language incredibly more complex than English in phonetics, syntax, and inflection. Perhaps the simpler languages are *less* primitive and language becomes less complicated with time? That is known to be true of English. It is probably true also of Chinese, long hailed as syntactically the simplest of all languages and claimed (not altogether correctly) to have no grammar at all.

Yet it is incredible that the first language could have been the most complex. Without pursuing this further, we must agree with A. Meillet, a great comparative linguist, that there are no primitive languages now extant and that no known idiom gives us any idea whatever of the characteristics of a primitive language.

Another bright idea is that since ontogeny (individual development) repeats phylogeny (evolutionary descent) we have only to study the acquisition of speech by children to learn how language originated. Largely abandoned but still advocated by an occasional diehard or eccentric, that theory has two flaws: (1) ontogeny does not repeat phylogeny, and (2) even if it did, that would not apply to individual and racial development of speech. The child is not repeating the acquisition of primitive language; he (whoever he may be) is learning an extremely sophisticated language. He is not doing this with a primitive brain just evolving the capacity for language, but with one already (even at his age) organized for learning and use of sophisticated language.

Perhaps we can at least determine when language arose by tracing the anatomical evolution of the vocal apparatus? That line is even now being followed seriously by some anatomists, but I think they are astray. A human brain in a monkey's body would probably mispronounce English words, but it would certainly produce a language. Anyway, Mann's myna could pronounce English words adequately with an extremely different vocal apparatus. In fact the phonetics do not matter: the mental capacity for language permits encoding and decoding information with any sounds whatever or, indeed, without sounds. Ideographs, already mentioned, are a language without specific reference to sounds, and there are several other soundless forms of true language.

Finally there is the brain, in which capacity for language is physically located. It has been argued that language will automatically emerge when a critical level of general intelligence or of brain size or complexity is reached. But the work of many students (including E. H. Lenneberg, who has also organized and summarized the evidence) has now shown beyond reasonable doubt that this is not true. Of course a certain level of general intelligence (if there is such a thing) is indicated, but it is the specific kind and not the quantity of brain that determines the capacity for language. That qualitative requirement has not been and probably cannot be determined from fossils. Brains do not fossilize, and despite much labor and many claims to the contrary it has not so far, at least, been possible to judge their language capability from their outer form as reflected in the skull cavity.

Some rather sterile lines of inquiry into the evolutionary aspects of language have been mentioned. There still are lines that have already proven fertile and that are prospectively even more so. We have established that not language *per se* but the capacity for language is an inherited genetic function of the human brain, and of no other among animals now living. The lines to follow are, then, clear: functional anatomy, physiology, genetics, and psychology of the human brain as related to all forms of linguistic behavior; close comparison of all these aspects with brains incapable of such behavior; and also further study of nonlinguistic communication in man and other animals with clear differentiation from truly linguistic communication.

CHAPTER NINE

The Future Evolution of Man

"The theory of natural selection can describe and explain phenomena with considerable precision, but it cannot make reliable predictions."
—ERNST MAYR

Everyone is aware, even from his personal experience, of great changes around us. Change, or, if you like, progress, is not only continuous; it also constantly increases in tempo. A European savant has fitted a curve to data for this acceleration and has concluded that at a finite time now not far in the future the rate of change will become infinite. The implication is that time itself will stop at that date. No alarmist, I do not give the date, and in any event it need not worry us. The savant's extrapolation is an extreme but otherwise rather typical example of how not to predict the human future. It adopts one of several quite different mathematically possible curves based on largely wrong, wholly inadequate data and assumes that this trend, and no other, will continue literally to the end of time.

It should be noticed, to begin with, that the changes in human life so visibly accelerating at this moment are mostly cultural, in the broad anthropological sense. They involve what we learn, make, and do, not what we are. Cro-Magnon man, who appeared in Europe some 35,000 years ago (give or take a few thousand), would look as normal in a space capsule as any astronaut and, given equivalent training, would doubtless manage just as well. To be sure, there have been some changes in head shape and in stature since then, but these have not been universal and probably are not wholly genetic. Proportionate numbers of races have changed, races have crossed and intercrossed, populations have moved—but that sort of thing has always gone on in the human ancestry, up to the most remote, and it has not recently changed the biological nature of the whole species *Homo sapiens* to a clearly significant degree. One unquestionably great and significant change is the enormous, now truly explosive increase in numbers in this species, but any eventual biological effect of this phenomenon is not yet evident.

Some students have come to believe that the biological evolution of man is at a standstill and that henceforth essential changes in our species will be cultural only. However, the consensus among biologists, at least, is that the bodily—somatic—evolution of *Homo sapiens* is continuing, for better or for worse, and will continue. There have been fewer than 2,000 generations since Cro-Magnon man appeared in Europe, a time too short for notable evolutionary change to be expected in a species under natural conditions. As will be specified further along, human populations have biological aspects that are known to lead eventually to marked evolutionary change in other species and must almost inevitably do so also in this one. Furthermore biology and culture are not two quite independent things such that one can, like a relay runner, pass on the torch of progress—or simply of change—to the other. They are parts of an interacting system, in which cultural change will certainly affect the nature and rate of somatic change and may well accelerate it. And cultural accomplishment has already reached the point where men could deliberately accelerate the evolution of the species and influence its direction.

The subject of man's bodily, somatic, and biological future evolution is therefore not idle. Its interdependence with cultural change must be kept in mind, but so far as practicable it is biological evolution, alone, that is here discussed. One must think first of a basis for prediction, and of course this involves knowledge of how, in fact, evolution occurs. Some articles of the less sophisticated Sunday-supplement type have assumed that effects of use and disuse are inherited and have predicted, for example, that the reluctance of the younger generation to walk when they can ride will result in reduction or loss of legs. In fact effects of use and disuse are not heritable. It is true that if legs were of no use at all (even on the pedals of a car) and if men were most attracted to girls with short, skinny legs or none, legs would probably be reduced or lost in some tens of thousands of generations. The present facts do not support such a prediction.

An approach that seems at first sight to be more sensible is to extrapolate an evolutionary trend into the future. Full reliance on this predictive method involves the theory of orthogenesis: that a trend once established continues indefinitely because of some inner urge, regardless of adaptation or outer circumstances. That, too, is now known to be false. Evolutionary trends, as long as they really last, have an adaptive basis, direct or indirect; they are governed by natural selection. They do not continue indefinitely. In due course they stop, change direction, or are even reversed. To extrapolate a trend into the future, one must be sure that the trend is really active at the present time and that its continuance will be favored by natural selection. Another, perhaps somewhat more sophisticated popular prediction has been that because man's brain and its bony case have greatly increased in size they will continue to do so—a figment accompanied by pictures of beings with enormous globular heads wobbling on tiny bodies. In fact it is unlikely that such a trend now exists. Cro-Magnon man had as large a brain as the average for living men of equal stature. Oddly enough, the still older Neanderthal man apparently had brains somewhat *above* the present average, a fact already known to Darwin and puzzling to him. The solution to the puzzle is that specialized Neanderthals were a side issue, not representative ancestors, and that moderate individual differences in brain size are not correlated with intelligence anyway. Still, these facts do speak against any currently active trend for increased head and brain size. Moreover, as will be further mentioned, there is little reason to think that a trend toward increased intelligence is now favored by natural selection, and the opposite may be true.

Other would-be predictive views on evolution involve a finality or a supernal goal. I will specify just one in order to show that their further omission here is intentional. The late Father Teilhard postulated an extreme orthogenetic, trend-bound course for evolution with man as a proximate goal and man's involution into God, Omega, the ultimate goal. There is no objective evidence for that frankly mystical view, and what is known of the course and the mechanisms of evolution is flatly inconsistent with Teilhard's postulate.

It is thoroughly established that natural selection is the essential orienting or directive factor in evolution. There is more to it than that, but reasonable prediction must at least depart from the basis of natural selection. Natural selection is a very complex process that is inadequately, in certain instances quite falsely, summed up in the phrase, "survival of the fittest." Reduced to barest essentials, natural selection is differential reproduction. It means that within a given, interbreeding population over the course of many generations individuals with certain genetic characteristics have more descendants than individuals with different characteristics. Hereditary factors of the former individuals will become more common in the population than those of the latter. That is the basic, most elementary process of directional, that is to say, nonrandom evolution.

Simple as it sounds, there are important subtleties in this process that are not generally known and that bear strongly on our own evolution and prospects. In the first place, natural selection does not directly favor fitness in the common sense of that word. What it favors is effective genetic controls for efficient reproduction over long periods of time. Usually fitness, in the vernacular sense of good health, strength, agility, intelligence, and so on, is correlated with efficient reproduction and hence is favored by natural selection, but that is not necessarily so. If it should happen, for instance, that weaker or less intelligent people have more descendants, then natural selection would lead to a decrease in strength or intelligence.

For practical purposes, the potential for natural selection in human populations can be subsumed in three broad categories: prenatal death, postnatal death, and fecundity. Prenatal mortality includes any factor that prevents a fertilized egg (a zygote) from developing into a baby born alive. The causes of such deaths are often, if not usually, genetic, and prenatal mortality thus has a definite selective effect. Since the genetic factors thus selected against are obviously and by definition disadvantageous, the result of this selection is advantageous. In the most technically advanced cultures some diminution of prenatal selection, for instance that caused by the Rh factor, has been achieved. On the whole, however, this form of selection still occurs in all human populations with little abatement.

Postnatal mortality selection occurs when a significant proportion of infants born alive fail to survive through their reproductive ages and when the survivors are, on an average, genetically different from the nonsurvivors. If few persons die before completing their reproductive roles, there is little possibility of natural selection on this basis. That condition is closely approached among the privileged members of affluent societies. It is estimated (by Kirk in a symposium cited in sources for this chapter) that 96 per cent of white American females born in 1960 will live to age 45. Only 86 per cent of those born in 1920 reached that age, but even that is a high percentage. In less privileged groups, including nonwhites in the United States, and in less affluent societies, postnatal mortality before completion of families is still quite significant and is a highly probable factor in continuing natural selection.

Fecundity selection occurs when some members of a population, genetically distinct in some way and as a broad average, contribute more descendants to later populations even though the mortality factor is nil or is discounted. Age of parents at childbearing affects numbers of descendants at later times and therefore has a potential for natural selection even among families with equal numbers of children, but with children born, on an average, at different parental ages. More obvious is the potential for selection in the fact that parents who live to complete their families may have families of quite different sizes. If, on an average, any 50 per cent of parents taken at random have 50 per cent of the offspring in a population (and have them at the same average parental ages), then there is no potential for fecundity selection. But if some group has more than its own proportion of children, there is a potential for selection, becoming stronger as the discrepancy in size of parent and children groups increases. This potential is present even in relatively simple conditions. For instance, in a recently split-off branch of the Navaho Indians, who are not primitive but who do lack many of the supposed blessings of urban civilization, 84½ per cent of the present population is descended from 48 per cent of the founders. In larger, more mobile and more sophisticated populations reliable data are harder to come by, but many clearly have even higher potentials for fecundity selection.

It appears that while mortality selection was being progressively relaxed among white Americans, fecundity selection was being maintained or increased. Recently there has been some evidence that fecundity selection has now begun to decrease in that relatively privileged and affluent group. It is, however, probable that it still affects even that group, and it seems to be potentially strong among American nonwhites. It demonstrably has high potential in some nontechnological or tribal societies, and probably has in countries less affluent and less industrialized than ours. It is also demonstrable that there is high potential for fecundity selection not only within but also as between human populations. If one community has a higher average birth rate than one genetically different from it, natural selection favors the former and the character of the species *Homo sapiens* as a whole, its total genetic pool, will be changed correspondingly.

Even in the most advanced communities there is thus still a potential for natural selection. What is actually happening? What in fact is being selected for and against? The sad, one might almost say the shameful, fact is that we do not know. There has been extensive discussion, much of it not only polemic but also abusive. There are few solid, unequivocal facts. To be sure, mortality selection, such of it as persists, is doubtless still in the direction of fitness, even in the vernacular sense. Genetic defects manifested before birth or in childhood certainly still reduce chances for reproduction and, unless wholly corrigible, continue to be selected against. Selection of that sort tends to preserve the *status quo* but presages no evolutionary advance.

Fecundity selection among different populations now tends to favor the relatively underdeveloped countries. In terms of continents, it now favors Latin Americans most, Asians next, then Africans, then (at some distance) North Americans, and finally Europeans least. Africans and Asians are still much affected by mortality selection, while North Americans and Europeans have minimal prereproductive mortality. Early mortality is likely to decrease in Africans and Asians. It therefore seems that the fecundity differential will probably continue and increase in favor of Africans and Asians, particularly. Hence one can venture a modest prediction: in the next few generations the average composition of *Homo sapiens* will probably change greatly in the ever-decreasing proportion of Europeans (as such and by descent). (It will be remembered that a large percentage of Latin Americans are non-Europeans or are vigorous hybrids with Europeans.) Since there is no reason at all to think that non-Europeans are on an average either less or more genetically desirable than Europeans, no prediction as to improvement or deterioration of the species follows from this probability.

Selection within populations is likely to be more important in the long run because it apparently has more potentiality for change to better or to worse than selection between populations. Unfortunately it is also more complicated, and our knowledge of it is extremely deficient in really clear and reliable data. Among white Americans, postnatal selection in general seems to be relaxing. Crow, probably the soundest authority on this point, holds that relaxation of selection is not likely in itself to cause much change in average gene frequency. There would, however, be more tendency for gene mutations to persist and to increase in frequency. As most mutations are disadvantageous (inadaptive), that would result in some genetic deterioration until either countered by renewed mortality selection or countered permanently by technological means, that is, by unrelaxing application of medical and sanitary measures for public health.

Although apparently now decreasing, the potential for total postnatal selection does still exist among white Americans and it is evidently stronger among nonwhite Americans. It is probably strong within most other populations of the world. Now, whether we like it or not, it is a fact that where concrete data on reproductive rates exist they usually show higher rates for those elements of the population that are lower by some criterion, or by several or all of such criteria as measures of intelligence, health, accomplishments, education, income, or social status. Within a given population it is thus probable, although it cannot be taken as fully established in a specified population, that there is potential selection for those elements of lower status by any of a number of criteria. But here is the rub, and hence comes acrimonious debate: selection is not actually occurring unless those of lower status are, on an average, *genetically* different from those of a higher status.

Many well-meaning but wishful-thinking commentators have declared by fiat that all persons of lower status are just underprivileged and not in any other sense inferior. That will not do as an answer to judicious inquiry. Intelligence, personality traits, talents, health, and other characteristics here involved are strongly affected by opportunity and environment but are also known to be influenced and limited by genetic inheritance. People are not genetically equal; some *are* conceived and born more intelligent, more healthy, and so on, than others. Certainly there are underprivileged people of low social status who are genetically just as capable and desirable as any of the highest status. But, even though unwelcome, it remains a definite possibility that the average genetic quality is lower on an average in persons of lower status and higher on an average in those of higher status.

This argument, of course, requires great qualification in connection with groups that are indeed and as a whole underprivileged for reasons that are not connected with their genetics. That is certainly true of nonwhites in the United States. It must, however, be added that selection, apparently strong, among the nonwhites themselves seems to favor those of lower status. If that adverse effect should be stronger in nonwhites than in whites—and we do not know whether this is true—the obvious remedy is to make nonwhites as privileged as whites. In any case, the fact that some people within a population are genetically superior or inferior to others has no modifying bearing on the political equality, rights, and dignity of all men and every man.

There is little persuasive evidence that fecundity selection has in fact lowered the intelligence, or other desirable genetic qualities, of a population. Data have been gathered too sparsely and over too short a time to hope for a clear answer. The statistics are tricky and, so far, unreliable. However, no honest, informed person can deny the *possibility* that selection is now favoring genetic deterioration in some human populations. That demands more investigation and consideration of countermeasures, even though the probability is still debatable.

We know that the human species could degenerate genetically and that it may be doing so right now. We also know that it could be improved genetically—who can believe that *Homo sapiens* is ideal just as it is?—although the possible limits of such improvement are unknown. These facts have long been recognized although not hitherto as well understood as now, which is not any too well! Proposals for improvement of the species have a long history, back into antiquity. Eugenics, as a more or less organized movement, became involved with racism and with erratic procedures such as compulsory sterilization based on inadequate or incorrect genetic grounds. Despite the ill fame so acquired, the eugenics movement survives innocuously as concern, which all of us should feel, for the quality of the genetic pool of the whole human species and desire to safeguard and improve it. An admirable outcome is genetic counseling, which consists largely of identifying particular genes (properly, alleles) with gross, undesirable effects and advising individuals or, especially, couples of the probabilities that these defects would appear in their descendants. The hope is, of course, that those for whom the probabilities are appreciable will forego having descendants.

That approach helps to eliminate some human anguish and so is certainly worthwhile. Let us hope that the basis of genetic counseling will become still broader and firmer and that it will be extended to the whole of mankind and not only, as at present, to a tiny minority, often of those who least need it. From an evolutionary point of view, such negative measures are desirable but not crucial. Another aspect of natural selection not generally understood or sufficiently emphasized is that it is usually and most strongly a stabilizing, normalizing influence, preventing or slowing and not hastening evolutionary change. Such selection results from the fact, a well-nigh universal fact among organisms, that under usual circumstances the extreme variants of a population are likely to have fewer descendants, an effect that may be produced both by differential mortality and by differential fecundity. Negative eugenic measures, seeking to eliminate genes (alleles) known or believed to be "bad," simply strengthen and accelerate the stabilizing natural selection that is active anyway, even in the most culturally complicated human populations.

The artificial reinforcement of stabilizing selection may in some degree come to compensate current cultural activities that tend to relax stabilizing selection, especially the medical treatment of genetic disabilities and increase in mutation-causing radiation. But no intensification of stabilizing selection is likely ever completely to eliminate genes considered undesirable, and no relaxation of that selection is likely ever to permit such genes to become universal or even usual. What is at stake is just the level of incidence of genetic defects in populations, which is of great medical and humanitarian concern but of little evolutionary importance. It does not, as a rule, change the norm or affect a direction of evolution. It should also be remembered that definition of a "bad" gene may be relative and changing, for a gene "bad" in some individuals and circumstances may be or become innocuous or positively "good" in others.

A positive eugenic approach would involve increasing the frequency of heredity that is desirable or, at least, desired. The problem of identification is here even greater, and a general solution is not even theoretically in sight. For one thing, in the present state of the art genes are usually identified by mutant forms (alleles) that produce abnormalities, not by their usual and desirable actions. For another, important human characteristics such as intelligence, talents, and general health depend on combinations of genes and interaction of total heredity, not on single genes that can be specified and promoted alone. Moreover, it would be extremely undesirable even if it were possible (which it is not) for all of mankind to be homogeneous, with a single set of "best" genes. There is no such set. Best for the *species* is an intricate mixture of different genes (alleles, again) many of which, taken individually and in particular circumstances, might be judged inferior to others.

Still a start could be based on the assumption that the gene sets of successful, productive, useful, intelligent people are likely to be superior, on an average, to those of people in similar biological populations who are failures, drones, parasites, and noddies. That is an assumption and it is one opposed by sentimentalists, but isn't it the most reasonable one? If so, then positive eugenic selection would involve long-range increase in the proportion of descendants of the superior group in the population. Several ways of accomplishing that end have been proposed, but none has been put into effect and, for various reasons, none is likely to be in the even vaguely predictable future. One, especially urged by H. J. Muller, is the preservation of sperm from eminent men for use after their death for artificial insemination of multiple women. That can probably be made technically feasible, but the social and psychological barriers (which nevertheless might conceivably change) are now completely prohibitive as regards application so general as to be effective.

Another possibility would be to apply social and governmental sanctions (for instance, in taxes and various forms of subsidies) so as to penalize large families of low status and reward those of high status. (The present tendency is usually exactly the opposite.) That is evidently impossible under present political conceptions, usages, and shibboleths. One surely should hope that it is compatible with some acceptable form of free and democratic society.

Most recent is the concept of genetic engineering: directed mutation, insertion of foreign but (presumably!) desirable genetic material into body cells, and eventually the synthesis and insertion of new, tailored genes. Not even the least of these things has ever been done with humans or with animals remotely similar to humans, and, despite some enthusiasts, there is little prospect of them in the near future. If or when there is any success along those lines, first will almost certainly be for the repair of genetically determined enzymatic deficiencies in particular organs, such as the liver. The effect would not be hereditary, and the evolutionary significance would be of no further importance but exactly the same as present medical treatment for some defects, such as the use of insulin to alleviate diabetes.

The problems of tailoring a gene and inserting it in human sperm or egg, making it hereditary, are so many and so little understood at present that reasonable prediction would place that in a future very remote indeed. Moreover, the human (or any other viable and natural) gene system is so intricately balanced that insertion of a foreign element, however well specified in itself, would probably have disastrous effects. Recognizing this, a few Cassandras (or are they Jeremiahs?) are already crying "Woe!" But even if it were not premature, the cry would probably be unnecessary. Mankind as a whole seems, not hopelessly but at present, both too inert and too ill-informed to take positive and effective collective action about our evolution as a species. However, those who learn that the individual action of inserting tailored genes produces abnormal children will surely have sense enough not to take that action. If, on the contrary, the results prove to be good and therefore are accepted, why cry "Woe!"? (I sometimes wonder whether this is the biologists' hopeful equivalent of the atomic bomb and a chance to share guilt feelings with the physicists.)

I have not made positive predictions. I have only mentioned and not discussed such obviously pertinent, indeed necessary topics as overpopulation, radiation and mutation loads, the balance of good and bad in medical and preventive procedures, and the possibilities of divergence or homogenization in the species. Predictions are being made, and those topics are being discussed. What I do hope to have done is to provide a basis for their comprehension and evaluation.

CHAPTER TEN

Biology and Ethics

"Optimists believe that ours is the best of all possible worlds. And pessimists are those who fear that the optimists are right." —T. DOBZHANSKY

Many biologists have considered ethics from the point of view of their science or (and this is not the same thing) have sought to derive ethics from biology. That is part of an even broader search, the search for naturalistic ethics, that is, an ethical or moral system rationally related to the nature of things or to the material universe. Different understandings of "nature," "natural," and "naturalistic" have led to much confusion and controversy.

There are, for a start, the term and the various concepts of natural law, often taken in a theological sense. I shall not discuss natural law at length because the theological aspects are outside my aim as well as my competence. Yet the subject must be mentioned because ideas of theological natural law have been related to those of scientific laws of nature and of philosophical naturalism.

Zoologists sometimes point out that guinea pigs are so called because they are not pigs and do not come from Guinea. Natural law, in a usual sense of the term, illustrates a similar caprice, for it is neither law nor natural. That sense, not the only one given the words, is that God issues edicts that are imposed upon phenomena.

Ethical imperatives or moral codes of supposedly divine origin may be given directly by revelation and inspiration. According to this view of natural law they may also be learned by reading God's mind through observation of His impositions on nature. But that is not natural in the usual sense and it is manifestly not naturalistic because it claims to be an approach to ethical principles of *supernatural* origin. The definition of naturalism pertinent to the present enquiry is precisely the opposite, the exclusion of any supernatural sources or sanctions for ethical systems.

Yet the theological idea of natural law as having an outcome in ethics has been carried over into concepts that are at least more nearly naturalistic. It has been thought that the laws of nature, as learned from strictly scientific enquiry, may have an ethical content. Even more broadly it has been felt that what objectively exists, being natural, is therefore in some sense right. In such crude form, these notions are clearly fallacious even though they cannot be entirely excluded from relationship with naturalistic ethics.

Such things as the law of gravitation or the equation $E = mc^2$ do not really have an ethical content. They only predict what will happen when a body falls or when a mass is converted into energy. At this level, which is that of scientific laws of nature, it is absurd to ask whether the law is good or bad. And yet what will happen in the specific case if, for example, we push an acquaintance out of a tenth-floor window, or if we convert mass into energy over a Japanese city—such things certainly do involve ethics. The fact that we can and must make such predictions does inevitably relate the laws of nature, themselves ethically neutral, to necessary ethical judgments. The point will be further developed later, but it should be noted here from the start how basic it is to a naturalistic ethic that men can predict or foresee the outcome of their voluntary actions and that they thereby become responsible for ethical evaluation of the results.

The naturalistic fallacy has a long history, generally dated from Hume in the eighteenth century. He pointed out that it is not logical to pass from "is" or "is not" to "ought" or "ought not." Another way of putting it is that description is not prescription. That really seems obvious enough, and the fallacy has been exposed so many times that another refutation is not necessary here. Yet the conviction that this is a fallacy is sometimes applied too broadly, and again what is truly fallacious sometimes slips by in more subtle ways. It is undoubtedly illogical to conclude that what *is* therefore *ought to be*. It is, however, equally illogical to make that the basis for a further conclusion that decision as to what ought to be cannot be based on consideration of what is—in other words, that naturalistic ethics are impossible.

Some of those who believe that science is (or should be) ethically neutral have fallen squarely into the naturalistic fallacy by their very efforts to avoid it. Insisting that they only describe and do not prescribe, they deny that any consideration of right or wrong applies to their subject matter. Yet in the sciences what is described often does in fact have an ethical aspect. From that point of view the "objective" scientist then runs some danger of condoning what is wrong or rejecting what is right. Or, at least, his attempted withholding of judgment may in itself amount to an ethical stand. A blatant but otherwise typical example was provided by the late A. C. Kinsey. He insisted on the scientific, putatively descriptive nature of his studies on human sexual activities. However, having (as he believed) established a certain prevalence of male homosexuality he called for legal reforms on that basis. This amounts to a decision that what *is* at least is not a social wrong. Kinsey was thus taking a definitely ethical stand in the name of scientific neutrality. His results have a bearing on ethical judgment, but the bearing is not that what is therefore should be licit. The social scientists can judge better than I, but I have the impression that similar or perhaps subtler examples, truly falling under the naturalistic fallacy, are quite common in their field.

(In pointing out the fallacy in Kinsey's work and in similar studies, I do not mean that the conclusions reached are necessarily wrong, but only that no logical basis for them has been given.)

Although scientists who intend to be naturalistic can thus wind up in the naturalistic fallacy through the back door, so to speak, it is an interesting observation that this fallacy is more likely to invalidate some nonnaturalistic theological views of natural law. If it were true that ethical prescriptions were determinable from supernatural impositions of law on phenomena, then that would indeed be a basis for believing that what is ought to be.

We have equal cause to reject the naturalistic fallacy and what I may call the counternaturalistic fallacy, that is, the belief that ethics cannot have a naturalistic basis but must have supernatural origins or sanctions. The fallacy of that point of view will be sufficiently demonstrated if we can supply the following two needs: (1) a naturalistic explanation of the fact that normal humans have a "moral sense" or feelings of ethical obligation, of values in terms of right and wrong, and (2) a naturalistic basis for making ethical decisions, for deciding which is better or worse of alternatives involving ethics.

Let us not be too dismayed if some attempts to set up naturalistic ethics turn out themselves to be fallacious, or if we cannot all agree as to either origin or criteria for naturalistic ethics. It would be enough to go on with if we could conclude that naturalistic ethics are eventually possible. In fact, I believe we can do considerably better than that.

Naturalistic ethics are now almost always considered in a biological and usually evolutionary context and will be so here, but in fact they were discussed and in some forms supported long before Darwin. Adam Smith held that the ethical sense arose in a natural manner, although he had no knowledge of biological evolution. Oddly enough, long after becoming an evolutionist Thomas Henry Huxley held the same view. He believed that the ethical (or moral) sense arose in a social context by imitation, sympathy, fear of disapproval, and hope for approbation. He believed this development to be not only independent of organic evolution but also decidedly opposed to it. His grandson Julian Huxley flatly rejects any conflict between organic evolution and ethics, and yet his views as to the origin of the moral sense are not profoundly unlike his grandfather's. They are more sophisticated and are expressed partly in Freudian terms with "superego" for "moral sense," but Sir Julian similarly sees the origin of ethics as what he calls a psychosocial phenomenon.

Waddington has carried somewhat similar ideas still further. He stresses the fact that man is an "authority acceptor" and an "ethicizing" animal, as had others before him in different terms. Waddington is critical of Julian Huxley and other predecessors for not discovering that "the function of ethicizing is to mediate the progress of human evolution, a progress which now takes place mainly in the social and psychological sphere." I may say that I see little real difference between this view and that of either of the Huxleys, and the difference from the eighteenth-century view of Adam Smith may not be entirely fundamental.

In the evolutionary context, the problem really becomes one of why and how organic evolution produced an animal capable of cultural evolution and of the ethicizing that helps to mediate cultural progress. There is no real doubt, and neither Julian Huxley nor Waddington has doubted, that the capacity or, one can say, the necessity for ethicizing is in fact a *biological* characteristic of the human species developed by natural selection because it is adaptive for that species. In generally biological but nontechnical terms, the direction of human adaptation early became one depending on individual flexibility with mainly learned abilities, with alternatives of action and necessary choice among them, with foresight as to the results of action, and with consequent responsibility for those actions. This kind of adaptation has always been carried out in a co-operative social milieu. It follows inevitably that men must and do learn, both from their fellows (parents, elders, priests, etc.) and from their own experience and introspection, to consider some actions "right" and some "wrong."

One must agree with Waddington that the acceptance of authority usually has been involved in the process. Nevertheless, it is not the only and probably is not even a necessary element in the human need for adaptive ethics. All this can be put in psychological and, indeed more narrowly, in psychoanalytical terms. Such terms are freely used by both Julian Huxley and Waddington and also to some extent by Dobzhansky, although he is less taken with Freudianism. I have not attempted a Freudian statement because I think the phenomena are at least as readily understood in biological terms —and I believe I understand the biological expression of them but doubt whether I fully understand the Freudian expression. It is, incidentally, interesting that Freud himself could not give a valid

Freudian expression of the biological aspects because his biology was outmoded even in his day and is now almost wholly rejected. He thought mostly in terms of recapitulation and of the inheritance of acquired characters, the formerly largely and the latter wholly discarded by modern biologists.

It is thus plausible and indeed, I think, practically certain that ethicizing, the capacity and the necessity for some system of ethics, arose in the course of human evolution in a completely natural way. That does not mean that men early and generally recognized the natural origin of the moral sense or that they adopted specific ethical systems that were naturalistic. We all know that they did not. Until quite recently virtually everyone who gave any thought at all to the subject believed that the moral sense was implanted, full-blown, by the Creator (in the Garden of Eden or elsewhere), and that the moral code followed was likewise of supernatural origin.

It is a peculiar fact that if those origins for the ethical sense and for particular ethics were sincerely believed, it commonly did not matter whether they were true or not. The results of loving your neighbor are much the same whether you consider this a commandment from God or just a good idea from an ecological point of view. We know that a great number of ideologically and theologically conflicting religious and ethical systems have all served adaptive biological and psychosocial functions with considerable success. Since they do conflict, some are necessarily false and in fact all may be. Thus, the little matter of whether they are true seems to have no pragmatic and perhaps even no ethical significance. However, from the point of view of our own day that possibly comforting thought is canceled out by two insuperable weaknesses.

In the first place, no given one of the systems of revealed or inspired religion and ethics is necessarily adaptive biologically or socially. They were not consciously adaptive in origin and they have been influenced, even manipulated, by followers who had quite other ends in view. There is thus no guarantee that they are or will continue to be adaptive. In fact some of the bitter conflicts between rival systems of supposedly revealed ethics are obviously inadaptive even when either system alone might be sufficiently adaptive for survival. Moreover, almost all of these ethical systems originated long ago under tribal, pastoral, or other primitive conditions. It is inevitable that they have become, to greater or less degree, inadaptive in the extremely different social and other environmental conditions of the present time. That has been compensated to some extent by the fact that the old systems now usually receive only lip service and in practice are largely replaced by pragmatic ethics of quite a different sort. However, this contrast between profession and practice itself introduces another source of stress, inadaptive both in individuals and in social groups.

The second major weakness in reliance on traditional ethical systems is that it becomes increasingly impossible to believe in their divine origin or to accept their ostensible basis as true. As noted previously, truth can be irrelevant to adaptive value if the untruth is believed. Also, psychologists have observed that many people can believe things that they know are not true. Yet there are many who value truth itself as an ethical good, who will not accept fiat as a substitute for reason, and who are not swayed by sanctions—such as those of heaven and hell—which they consider mythical. Ethical systems that are inacceptable to large segments of society are inadaptive to some degree for that reason alone if for no other.

The traditional systems have not become wholly inadaptive—otherwise they or those who maintain them would have become extinct. Nevertheless, they manifestly have inadaptive aspects, which are becoming increasingly crucial, and which cannot continue indefinitely in adaptive and adapting societies. If they are simply rejected, that reduces them to completely inadaptive status and demands their replacement. It may be one of Freud's greatest contributions that he insisted on both these points: first, that a supernaturalistic basis for ethics is no longer adaptive, and second, that an ethical system on a rational (or naturalistic) basis still is a necessity. He used other words, for example:

> The ethical commands to which religion seeks to lend its weight require some other foundations instead, for human society cannot do

without them, and it is dangerous to link up obedience to them with religious belief.

We have, then, this situation: Man is an ethical animal; in fact he is *the* ethical animal. His ethicizing is a biological adaptation necessary to success in his way of life. Supernaturalistic ethics once universal and still usual give no promise of continued adaptation, biological or psychosocial, but on the contrary seem to be increasingly inadaptive. Therefore, continued human welfare requires ethical systems that are not supernaturalistic but are naturalistic.

Naturalistic ethics are reduced to the simplest, most widespread, and most banal form in the ethic of the golden rule and the ethic of social conformity. Supernatural origin has been claimed, too, for these ethics, but that is not necessary. They are sufficiently validated in a naturalistic way by the fact that they usually work. Yet these pragmatic ethics alone are hardly adequate for leading a good life, and they are not invariably adaptive. Conformity, although usually adaptive in a given situation, reduces adaptability under inevitable changes in the social milieu and hence has the same inadaptive potential as biological overspecialization or lack of variation. Moreover, the golden rule and social conformity do not meet all of life's ethical problems, and they often conflict with each other. The soldier conforming in war is doing as others would like to do to him rather than as he would like them to.

More extensive grounds for naturalistic ethics were already sought by John Locke in the pleasure principle, carried to greater extreme by Jeremy Bentham. Pleasure is good and pain is bad, but even in the extended form of the greatest pleasure for the greatest number this leads to ethical anomalies and contradictions. Altruistic behavior, which may be good by any yardstick, even a strictly biological one, is not overtly pleasurable. With more modern sophistication one can, of course, point out that pain may itself give pleasure, but a simple ethic of pleasure and pain cannot now be considered adequate, if acceptable at all.

Since Darwin the search for naturalistic ethics has generally taken some account of evolution, and rightly so because man and his need for ethicizing are results of the evolutionary process. Nevertheless, ethical systems proposed as evolutionary have frequently been fallacious. I shall mention some, even of those manifestly inacceptable, to exemplify their variety without taking space here to expound or refute all of them at length.

An early conclusion supposedly drawn from Darwinian evolution was that natural selection, "survival of the fittest," is a struggle without rules from which not only the fittest but also the best in an ethical sense emerge victorious. That idea became involved not only in capitalistic *laissez faire* but also in Marxian dialectic. It is fallacious even in its supposed biological basis, because natural selection does not necessarily or even usually act in the postulated way. It has not acted that way in the specific processes by which ethicizing man arose. Even if that were not true, there would be no reason to believe that the ethic of struggle would be adaptive for man as of now. That was the point that Thomas Henry Huxley made forcefully and at length in 1893, concluding that human ethics are opposed to, not derivable from, the then usual concept of the struggle for survival. It has been pointed out that Huxley, the agnostic, accepted, apparently intuitively, ethics indistinguishable from those maintained by the Church of England in the Victorian era. That is true and amusing, but Huxley's main point, that (in later jargon) ethics are species-specific, was well taken and profound and has been unwisely ignored by many of his successors.

In more general and less brutal terms the ethic of struggle can be replaced by the ethic of survival, for example, by Leake. We may agree that survival seems a rather good idea, but as an ethic this is inadequate. To be sure, without survival no other good can follow for the species in question. Since individuals never survive, the ethic applies only insofar as their actions may promote survival of the species. Such activity is indeed biologically adaptive and a naturalistic good, but its application is not simple. Unbounded procreation might seem to be most likely to promote survival of the species, but the evolutionary facts are that this is not so for many species and particularly not so for our species. Further, the pleasure-pain principle is applicable at least to the extent that mere survival accompanied by general misery cannot be considered ethically desirable. And still further, no known actions of man can guarantee or even make probable the *indefinite* survival of our species, which will almost certainly become extinct in due course whatever we do now.

Another approach is through the concept of society as a supraorganism, as formerly expounded, for example, by Gerard. This idea is not necessarily evolutionary and was indeed already discussed by Comte. In the evolutionary context, increasing integration of the supraorganism has been considered a progressive trend. If, then, the supraorganism, that is, a society or a nation, is an entity distinct from and superior to the individuals composing it, it is ethical to sacrifice any or all merely individual goods to the good of the state. That idea is all too familiar to us, and its several fallacies are doubtless equally familiar outside of the "organic states" of Fascists and Communists. One of the biological fallacies is that the organism-supraorganism comparison is an imperfect analogy and that the part on which that conclusion depends is quite simply false. Further, the comparison of human and insect societies, often used to support the supraorganismic analogy, is largely invalid because the biological basis of socialization is quite different in the two cases. The ethics of the ant are fit only for an ant. Alfred Emerson knows as much about insect societies and has studied the supraorganism analogy as profoundly as anyone, and he has pointed out the unreliability of these analogies.

There are numerous biological phenomena that have been considered general tendencies of evolution and have been involved in proposals of evolutionary ethics. I merely list a few and name a representative authority who has discussed or endorsed it. (References to some of their works are given hereafter in "Sources and Second Thoughts.")

Increase in negentropy or decrease in entropy (Patten)
Maximization of metabolism (Lotka)
Minimization of effort (Zipf)
Increase in homeostasis, individual or social (Emerson; Lerner)
Increase in complexity (almost any text on evolution)

It is doubtful whether all, or perhaps any, of these are really *general* trends in the history of life, although with the probable exception of minimization of effort all seem to have occurred at one time or another in the human ancestry. The importance and roles of these various factors in evolution over-all have little relevance to the present topic unless it can be shown that there is some connection with ethics, that the "is" or "may be" of these phenomena can conceivably be read as "ought." I do not believe that there is an "ought" inherent in any of these processes. They may have ethical bearings not in themselves but as means to ends. For example, biological or social homeostasis maintains stability and normal function in populations that are well adapted to existing conditions. As long as these conditions obtain, actions that promote or increase homeostasis may be desirable both in a material and in an ethical sense. However, the ethical "good" is not the homeostasis in itself but the maintenance of adaptation by its means. Under changing conditions the resistance to change of homeostatic mechanisms makes the effects of homeostasis undesirable, at least, if not unethical. Thus one can hardly say that homeostasis, negentropy, minimum effort, and so on, are good or bad *per se*, but only that they have effects that may be either good or bad depending on circumstances and outcomes. As with the law of gravity, the moral issue is not in the general principle but in human involvement with its consequences in particular instances.

A form commonly taken by proposals of evolutionary ethics is what may be called "trend" ethics. An over-all evolutionary trend is inferred or postulated. It is then concluded that the continuance of that trend is desirable and that whatever promotes it is ethical. Although that is far too limited and elementary a statement of his views, trend ethics have been particularly expounded and supported by Julian Huxley. His conclusion as to evolutionary ethics, "in the broadest possible terms," is as follows:

> That it is right to realize ever new possibilities in evolution, notably those which are valued for their own sake; that it is right both to respect human individuality and to encourage its fullest development; that it is right to construct a mechanism for further social evolution

which shall satisfy these prior conditions as fully, efficiently, and as rapidly as possible.

Huxley goes on to point out that translation of these generalities into concrete terms "is a task for an entire generation"—for many generations at the present rate, I would say. It is also implicit here and partly explicit elsewhere in Huxley's work that his general, dynamic evolutionary ethic requires supplemental ethical principles. That is true even in the larger picture. It is especially true in connection with daily and particular exigencies, which may seem less important than the grand sweep of sociobiological change through all ages to come but which actually raise the ethical problems that are major to us as individual humans.

Although we would supplement them and might express them somewhat differently, most of us would agree with Huxley's main conclusions. The fact that we do agree that Huxley's ethics are ethical makes it the more imperative for us to examine their derivation and, if need be, seek remedies for any weaknesses found.

Huxley's argument is based on evolutionary trend. He maintains that, with all its multifariousness and its innumerable side branches or dead ends, the history of life has evinced one direction legitimately called progress. This trend has included heightened organization, complexity, and homeostasis. My previous conclusions that these phenomena or principles are neither ethical nor unethical in themselves would apply here and would tend to refute Huxley's conclusion if it were not that he subsumes the *results* under a broader principle of improvement:

> In the organic phase of evolution, this progress depends on all-round improvement as opposed to the limited improvement or one-sided specialization which, it can be demonstrated, automatically leads sooner or later to a dead end.

"Improvement" is an evaluating word. It means "change for the better," so that when unlimited improvement is taken as the criterion for the ethically good, what is being said is simply that it is good to become better. The argument becomes circular to the point of meaninglessness unless some external, independent criterion of "better" is found and applied. It is surprising and, at first sight, dismaying to find Huxley maintaining that the concepts of progress, of higher and lower, of better and worse, "spring automatically to the mind" and are known "as an immediate and obvious fact." If, after all, we are to take intuitive concepts as ethical principles, then purported derivation of naturalistic ethics from the facts of evolution is either irrelevant or spurious.

Fortunately Huxley does not leave the matter at that. Huxley provides criteria for improvement by exemplifying it, notably as "evolution towards greater capacities for feeling, knowing, willing, and understanding." Such is by no means the only possible definition of improvement. Tapeworms improved their chances for survival when they lost their digestive apparatus and hypertrophied their reproductive organs. Yet it is reasonable to consider capacities for feeling, knowing, willing, and understanding as improvements, and if that highly restricted definition is agreed upon, the matter can be discussed clearly in those terms. However, this concept of improvement and of evolutionary progress is strictly *ad hoc*. Among all the millions of living species it is fully relevant to only one: *Homo sapiens*. It may also apply in part and with rapid dilution to a few of man's closer or more distant relatives among the higher vertebrates. Improvement, so defined, is obviously absent or has reached only extremely low levels among the vast majority of evolutionary lines. It is not a general tendency in evolution, but a tendency limited to a very small minority of evolving organisms and fairly continual in late phases of the ancestry of a single one only.

An ethic so derived is not based on the nature of evolution but on the nature of man. It is all the better for that. Interestingly, in this analysis of Julian Huxley's thought we arrive very nearly where Thomas Henry Huxley started, although of course his expression of the matter now seems incomplete and indeed rather primitive. The agreement is that a workable naturalistic ethic is derivable not from the evolutionary process in general but from the evolutionary status of man in particular.

That is not to say that the evolutionary process in general is irrelevant and that in seeking a naturalistic ethical system we wind up with one that is nonevolutionary, as T. H. Huxley was inclined to think. The evolutionary process in itself is nonethical—there simply is no point in considering whether it is good, bad, a mixture of the two, or neither. As would be expected, some mammals other than man do seem to have slight rudiments of a moral sense, but on the whole it is nonsensical to speak of ethics in connection with any animal other than man, and still more so in connection with the plant kingdom—which after all is just as much a part of evolutionary history and trends as the animal kingdom. No, *man* is the only ethical organism in the full meaning of the words, and there are no relevant ethics except human ones. But man, with his ethical sense, was produced by evolution as the result of one evolutionary sequence. Human evolution continues and man is subject to all evolutionary principles, not simply as generalities but as they apply in his specific and extremely special circumstances. A rational, naturalistic system of ethics cannot be independent of evolution, but neither can it be derived from evolution as a general or merely abstract principle.

C. H. Waddington is another biologist who has developed an evolutionary ethics contemporaneously with Julian Huxley and in interaction with him. Waddington is in general agreement with the content and also the basis of Huxley's ethical proposals, but he observes some circularity in Huxley's reasoning, in part as I have already mentioned. He believes that this circularity is removed by understanding that "the existence of ethical beliefs is a necessary part of the evolutionary system." Waddington has made a great contribution insofar as he has expounded the evolutionary origins and function of "ethicizing" or of the moral sense, a subject implicit but not always so fully explicit in previous studies. I have referred to that earlier in this discussion and need not expand on it here.

Waddington divides the study of ethics into three parts or phases: (1) the development and function of a moral or ethicizing sense, as just noted; (2) the acquisition of a particular ethical system, and (3) the judgment of alternative ethical systems by superior or external criteria.

His third point or approach must of course determine any recommendation he might himself make as to the second point, the acceptance of a specific ethic. Here I find him obscure and contradictory. His metaethics, as I prefer to label what he calls "wisdom," have the following basis:

> I argue that the observation of the world of living things reveals a general evolutionary direction, which has a philosophical status similar to that of healthy growth, in that both are manifestations of the immanent properties of the objective world. Finally, I conclude that any particular set of ethical beliefs, which some individual man may advance, can be meaningfully judged according to their efficacy in furthering this general evolutionary direction.

Except for the remark on immanent properties, this is virtually identical with Julian Huxley's metaethics, with the same partial validity and the same weaknesses, already adequately noted. That the "general evolutionary direction," if there were one, would be a manifestation of "immanent properties" seems to me hardly more relevant in an *ethical* context than that the fall of a stone is a manifestation of the immanent property of gravitation. It is further noteworthy that cancerous growth is as much a manifestation of nature as healthy growth, and that by far the commonest trends in evolution are those that lead to extinction.

At a later point in his discussion, Waddington expounds a metaethical principle that contradicts the one quoted above, and here I can entirely agree:

> The problems of the organization of the human species are *sui generis;* very little help concerning them can be found by looking either at evolving animal species or developing animal individuals.

The contradiction is only partial, for Waddington applies his second metaethic within and as an adjunct to his first:

> The general anagenesis [supposed progressive trend] of evolution is towards what may crudely be called richness of experience. In the human evolutionary system, progress in this direction demands the formation of mental systems having some degree of authority. . . .

The first sentence essentially expresses an *ad hoc* concept of improvement like Huxley's. The second expresses Waddington's emphasis on the function and need for ethicizing by means of authority acceptance in the human species. As Waddington notes, the application of this general principle to concrete ethical problems is not simple or obvious. In taking improvement of authority acceptance as an ethical principle there is confusion of means and ends. There is also the danger of relating this ethic in a favorable way to authoritarianism.

I have tended to stress what I consider questionable or, at least, weak points in the search for naturalistic ethics in an evolutionary context. That seems most interesting and at this time probably most useful. However, I should also stress that, as I believe, most evolutionists agree with Huxley as to what are desirable naturalistic ethics specifically for man and with Waddington as to the evolutionary origin and function of the moral sense. Disagreement is mainly as to whether those conclusions really arise from or are validated by any general tendency of the evolutionary process. Dobzhansky, another evolutionary biologist who has deeply studied the nature of man, biological and social, has pointed out that even trends that really did occur in the specific human ancestry are not necessarily good guides to evolutionary ethics. For example, in primitive man there were certainly the beginnings of splitting not only into races (which need not and should not be antagonistic) but also into distinct and inevitably antagonistic species. It is also clear that significant increase in average human intelligence stopped long ago—tens of thousands of years, at least—and although not clear it is at least possible that an opposite trend is setting in. Few people would consider either of those developments ethically desirable, and their undesirability can be argued on biological as well as other grounds.

Biological trends are almost entirely directed by natural selection. Social trends operate within the framework and limitations of the biological factors and represent a remarkably flexible form of biological adaptation. I cannot here discuss the nature of these two factors in human evolution or the relationship between them. For that I can best refer you to Dobzhansky's books and also to some works of mine, cited hereafter. The point pertinent here, adequately supported elsewhere, is that neither strictly organic nor social evolution necessarily leads to improvement in any sense of the word acceptable for the human situation. In fact either, in their ways that *in themselves* are blindly amoral, may have results extremely undesirable for mankind. But here is the saving grace: both organic and social evolution are now to some degree, even though a limited degree, within our own control. We, alone among all organisms, are aware of our own sociobiological evolution, can judge what seems to us desirable or undesirable in that evolution, and can deliberately work for the desirable.

If ethics could be reduced to ethicizing, authority acceptance, and nothing else, then such discussions as this would be a mockery. Unless we reject naturalism entirely, that would leave us firmly impaled on the naturalistic fallacy that what is ought to be. There is really no point in discussing ethics, indeed one might say that the concept of ethics is meaningless, unless the following conditions exist: (*a*) There are alternative modes of action; (*b*) man is capable of judging the alternatives in ethical terms; and (*c*) he is free to choose what he judges to be ethically good. Beyond that, it bears repeating that the evolutionary functioning of ethics depends on man's capacity, unique at least in degree, of predicting the results of his actions. A system of naturalistic ethics then demands acceptance of individual responsibility for those results, and this in fact is the basis for the origin and function of the moral sense.

There are both thesis and antithesis in such statements. As an authority acceptor, any individual obviously does acquire much or all of his ethical system from his society, represented first by a parent, later by an ever-expanding social milieu. That has often been taken as a basis for shifting responsibility from the concrete individual to that comfortably abstract entity "society." Although a trace of truth can be involved, on the face of it the height of contradiction is achieved when society is blamed for an individual's being antisocial. (The trace of truth can derive from society's having presented so unethical an aspect to the individual that he is *right* to be antisocial.) Wherever and however the individual acquired a specific ethic, and even if he acquired none at all, it remains a fact, and even a biological fact, that he is responsible for his actions. (To discuss the special cases of abnormal individuals biologically incapable of foresight would take us too far afield here.)

This discussion has been directed to certain of the very broadest aspects, only, of its subject. It has entirely neglected myriad questions as to what, in detail and in particular instances, could be judged naturalistically ethical behavior. It has indeed neglected some broad aspects. Two more of these may be mentioned, each so briefly as merely to indicate the nature of the problems of principles.

First is the ethical distinction between the welfare of the individual and that of the social group. This has already been mentioned in the form of the supraorganism analogy, shown to be fallacious in this application. A more instructive analogy is in the operation of natural selection, which usually promotes the immediate welfare both of the group (population) and of its constituent individuals, but which may sacrifice the welfare of some individuals to that of the group and may, more rarely, favor individuals at the expense of the group. That analogy is also imperfect, but it is true in human societies also that group welfare and individual welfare commonly coincide and reinforce each other, but that there are many instances of conflict of one with the other. An ethical decision frequently requires a compromise between the two. This is related to the long evolutionary history of individualization and socialization. It is worth noting that insect socialization was achieved at the *expense* of individualization but that human socialization arose *by means* of individualization.

The second topic to be touched in passing is that a single, absolute system of ethics for all of mankind would almost certainly be disadvantageous—indeed itself unethical—in the long run. A system of ethics is functional in human sociobiological evolution to the extent that it is adaptive to existing conditions. Now, the human condition has changed radically in the past, is now changing at a rate fantastically higher than ever before, and will certainly continue to change in the future. To remain functional, ethical systems surely must also change in some respects although not in all. Here another biological analogy is pertinent and valid. Maintenance of organic adaptation by natural selection under changing conditions is possible only if there is genetic variation in the population. It is also true that adaptive change in ethical systems is possible only if they vary. Variation in ethical standards is desirable and is probably necessary for survival in the long run. In both cases, however, it is wise to remember that variation must not affect the minimal agreement required for the organic or the ethical system to be viable in the existing environment.

Semifinally, I wish to quote in somewhat shortened and altered form a passage from an earlier work of mine which Dobzhansky also chose to quote at the end of one of his books:

> Man has risen, not fallen. He can choose to develop his capacities as the highest animal and to try to rise still farther, or he can choose otherwise. The choice is his responsibility, and his alone. . . . Evolution has no purpose; man must supply this for himself. . . . It is futile to search for an absolute ethical criterion retroactively in what occurred before ethics themselves evolved. . . . The new [sociobiological] evolution [peculiar to man] involves knowledge, including the knowledge of good and evil.

Finally, after writing at such great length I find that Mark Twain said much of this in a more succinct way:

"Man is the Only Animal that blushes. Or needs to."

Sources and Second Thoughts

CHAPTER ONE

This chapter is based on an address given at the first degree-conferring exercises of York University, Toronto, Canada, and the dedication of new science facilities there. A written version, substantially as here, was printed in *The American Scholar*, Vol. XXXVI, No. 3 (1967), pp. 363-377.

It is not necessary and it would be invidious to name the young instructors, the old professor, and a few others whose views are mentioned with disapproval. They are real and worthy individuals, but I would not state their views if these were not shared by a number of other biologists. I will name the biologist who thinks that life is the secret of DNA, and not the other way around, because I heartily endorse that view and because in his case I think "stormy petrel" is a compliment: he is Barry Commoner.

Here and hereafter many readers may want to know more about the breakthrough in molecular biology and especially in studies of DNA. The best book for that purpose is:

Watson, James D. *Molecular Biology of the Gene* (New York and Amsterdam: W. A. Benjamin, 1965).

A simpler book, slightly disappointing in spots and with a title that annoys me, is nevertheless equally authoritative and has the further virtue of making more connection between molecular and organismal levels:

Beadle, George and Muriel. *The Language of Life* (Garden City: Doubleday, 1966).

Things are still moving so fast in this field that every book about molecular biology is at least slightly out-of-date when published.

This book will be, too, but just by way of example I cite what is, on the day this sentence is written, the most recent word on the genetic code:

Garen, Alan. "Sense and Nonsense in the Genetic Code." *Science*, Vol. CLX (1968), pp. 149-159.

The evolution of biomolecules, barely mentioned here, has been discussed at some length in Chapter Three. By the way, the term "biomolecule" is to be preferred to "organic molecule" or "organic compound" when reference to compounds synthesized by organisms is intended. "Organic" in the chemical sense has come to be applied confusingly to any molecule including carbon, and many of these have nothing to do with organisms. A macromolecule is, of course, one that is big as molecules go. The largest macromolecules are biomolecules, and they are the most interesting biomolecules, but there are plenty of small biomolecules.

CHAPTER TWO

The original version of this chapter was written as a contribution to "International Colloquium II: Biology-History-Philosophy" at the University of Denver in November 1967. A somewhat curtailed and slightly garbled version was published in the *University of Denver Magazine,* Vol. 5, No. 3 (Jan.-Feb. 1968), pp. 21-25.

Here again I prefer not to name the professor whose statement I can only qualify as utter nonsense. I can assure the reader that he is a real person, not usually nonsensical, and that he really did write the nonsense here quoted. The exclamation in ordinary parentheses (not square brackets) is his. The quotation from E. W. Sinnott is from *The Biology of the Spirit* (New York: Viking, 1955). A more extended discussion of that book, written while my friend Sinnott was alive (alas! he is no longer) is in Chapter Eleven of my book *This View of Life* (New York: Harcourt, Brace & World, 1964).

There are a few moderately technical terms in this and some other chapters, but I think they will be familiar to most readers. The biochemical and genetical terms at this level are all fully explained in Simpson and Beck, *Life: An Introduction to Biology*, 2d ed. (New York: Harcourt, Brace & World, 1965).

It is now reasonably well established that all but three of the triplet codons do code for one of the amino acids, there being from one to six codons for each of twenty amino acids to a total of 61 codon triplets. The other three are called "nonsense" triplets just because they do not code for an amino acid, but they may well have some other function. They may, for example, serve as figurative punctuation in the message, indicating where specification of a protein ends, although this is considered improbable by some biochemists. In any case, it is premature to conclude that they literally make no sense.

It is the existence of more than one codon for the same amino acid that is oddly called "degeneracy" by the communication theorists. I do not know why the obvious term "redundancy" or simply "repetition" is not used instead. In fact, it is quite easy to hypothesize reasons why such "degeneracy" may be functional or even essential for optimal adaptation, although such utility has not yet been clearly established. This is another example of prejudging that something in evolution makes no sense, which is extremely unlikely and which is stultifying for further fruitful research.

The books by Watson and by the Beadles cited for the preceding chapter are equally or even more pertinent here.

CHAPTER THREE

A first version of this chapter was written to be presented at a colloquium on "Protides of the Biological Fluids" in Bruges, Belgium, in the summer of 1964, but I was unable to attend the colloquium. The paper was partly revised and rewritten as an essay published in *Science*, Vol. CXLVI, No. 3651 (1964), pp. 1535-1538. That version is incorporated here with some editorial changes.

The following are references to authorities and studies directly cited in the text:

Anfinsen, C. B. *The Molecular Basis of Evolution* (New York: John Wiley & Sons, 1959).

Florkin, M. *Biochemical Evolution* (New York: Academic Press, 1949).

Goodman, M. "Man's Place in the Phylogeny of the Primates as Reflected in Serum Proteins." In *Classification and Human Evolution,* S. L. Washburn, Ed. (New York: Wenner-Gren Foundation, 1963), pp. 204-234.

Hill, R. L., J. Buettner-Janusch, and V. Buettner-Janusch. "Evolution of Hemoglobin in Primates." *Proc. Nat. Acad. Sci.*, Vol. L (1963), pp. 885-893.

Margoliash, E. "Primary Structure and Evolution of Cytochrome c." *Proc. Nat. Acad. Sci.*, Vol. L (1963), pp. 672.

Mora, P. T. "Urge and Molecular Biology." *Nature,* Vol. CXCIX (1963), pp. 212-219.

Pontecorvo, G. "Microbial Genetics: Retrospect and Prospect." *Proc. Roy. Soc. London*, Vol. B-CLVIII (1963), pp. 1-23.

Sanger, F. Quoted and cited in C. B. Anfinsen, 1959.

Sonneborn, T. M. "Implications of the New Genetics for Biology and Man." A.I.B.S. Bull. (April 1963), pp. 22-26.

Wald, O. "Phylogeny and Ontogeny at the Molecular Level." In *Evolutionary Biology*, A. I. Oparin, Ed. (London: Pergamon, 1963), pp. 12-51.

Weiss, P. "From Cell to Molecule." In *The Molecular Control of Cellular Activity*, J. M. Allen, Ed. (New York: McGraw-Hill, 1961), pp. 1-72.

Williams, C. A., Jr., and C. T. Wemyss. "Experimental and Evolutionary Significance of Similarities among Serum Protein Antigens of Man and Lower Primates." *Ann. N.Y. Acad. Sci.*, Vol. XCIV (1961), pp. 77-92.

Zuckerkandl, E. "Perspectives in Molecular Anthropology." In *Classification and Human Evolution,* S. L. Washburn, Ed. (New York: Wenner-Gren Foundation, 1963), pp. 243-272.

Zuckerkandl, E., and L. Pauling. "Molecular Disease, Evolution and Genetic Heterogeneity." In *Horizons in Biochemistry*, M. Kasha and B. Pullman, Eds. (New York: Academic Press, 1962), pp. 189-225.

At the present writing (in May 1968) it is less than four years since the essay on which this chapter is based was written. The study of biomolecular evolution has accelerated markedly in that time, and it is becoming one of the more active fronts of research in biology. There is great temptation to expand this chapter, but it would then become a book in itself. This chapter is concerned with attitudes and principles and not with specific findings on the evolution of particular biomolecules. I do not think that its general thesis requires more modification than I have given it in the light of research and publications of the last three years or so. I do supplement it with the following small sampling of recent literature:

Florkin, M. *A Molecular Approach to Phylogeny* (Amsterdam, London, New York: American Elsevier Pub. Co., 1966).

Hafleigh, A. S., and C. A. Williams, Jr. "Antigenic Correspondence of Serum Albumins among the Primates." *Science*, Vol. CLI (1966), pp. 1530-1535.

Jukes, T. H. *Molecules and Evolution* (New York: Columbia University Press, 1966).

Neurath, H., K. A. Walsh, and W. P. Winter. "Evolution of Structure and Function of Proteases." *Science*, Vol. CLVIII (1967), pp. 1638-1644.

Sarich, V. M., and A. C. Wilson. "Rates of Albumin Evolution in Primates." *Proc. Nat. Acad. Sci.*, Vol. LVIII (1967), pp. 142-148.

Sibley, P. G. "Proteins: History Books of Evolution." *Discovery*, Vol. III (1968), pp. 1-20.

Zuckerkandl, E. "The Evolution of Hemoglobin." *Sci. Amer.*, Vol. CCXII, No. 5 (1965), pp. 110-118.

Reference is made in this chapter to general causative theories of evolution and specifically to the synthetic theory, supported by a majority of modern biologists. The following is a small selection of works on that subject in English. These references are relevant throughout the present book.

Darwin, Charles. *On the Origin of Species. A Facsimile of the First Edition with an Introduction by Ernst Mayr* (Cambridge: Harvard University Press, 1964).

Dobzhansky, T. *Mankind Evolving* (New Haven: Yale University Press, 1962).

Grant, V. *The Origin of Adaptations* (New York: Columbia University Press, 1963).

Huxley, J. *Evolution: The Modern Synthesis.* 2d ed. (London: Allen & Unwin, 1963).

Mayr, E. *Animal Species and Evolution* (Cambridge: Harvard University Press, 1963).

Simpson, G. G. *The Meaning of Evolution.* Rev. Ed. (New Haven: Yale University Press, 1967).

——. *This View of Life* (New York: Harcourt, Brace & World, 1964).

Tax, S., Ed. *Evolution after Darwin.* Three vols. (Chicago: University of Chicago Press, 1960).

CHAPTER FOUR

A version of this chapter was given as an address at the Homecoming Weekend Convocation of Dickinson College, Carlisle, Pennsylvania. It was published in the *American Scientist*, Vol. LV, No. 2 (1967), pp. 161-175, and is incorporated here without essential change.

References to Benjamin Rush, who was a founder of the college, and to Fuller Spencerton Baird, who was both a student and a teacher there, are specific to Dickinson College. Those specific references are nevertheless retained in this more general work because Rush and Baird are excellent symbols of two aspects of biology: technological or applied, and research or pure science, respectively.

When this chapter was published as a journal article, the remarks on predator control and tax-supported extermination of wildlife elicited a strong reaction from the Bureau of Sport Fisheries and Wildlife, Fish and Wildlife Service, U. S. Department of the Interior. On 13 April 1967, which was after the periodical version of this chapter was written, the Secretary of the Interior approved a new statement on animal damage control policy. The proem of that lengthy statement says: "The Bureau's animal damage control program will be designed in a manner which will ensure the maintenance of the varied native wildlife and wildlife habitats of the

United States." Damage envisioned as requiring control includes transmission of human disease, threats to human safety, economic losses in urban and industrial areas, disruptions of forest and range management, damage to crops, predation on livestock, and communication of disease to livestock. Regarding predators, there is an explicit statement that bears and "lions" (meaning mountain lions or pumas) "are particularly prized parts of the Nation's wildlife heritage." If it is necessary to "remove" them (in almost all instances for "remove" read "kill') because they are damaging or believed to be "about to do damage [!] to livestock or natural resources," a written justification must be prepared. However, such justification may be submitted *after* the "removal."

All of us must welcome a statement explicitly recognizing the value of wildlife and specifying, as this does, that esthetic values are included and not only the pleasure some people derive from killing. However, I cannot avoid the feeling that this is in some part, perhaps even in greatest part, lip service. The list of situations requiring "control" hardly suggests radical advance over a policy to kill off all animals contrary to man's narrower interests in any way or degree, or believed by anyone to be so. A personal communication informed me that it is not Bureau policy to exterminate any species, even locally. However, that is not in the printed policy statement (*Man and Wildlife*, U.S. Dept. of the Interior, 12 May 1967), which, on the contrary, explicitly states that "social and economic benefits may . . . offset . . . the loss of the controlled species." Moreover, it is evident to anyone who knows government hunters, trappers, and poisoners in the field that their personal aim is generally extermination, whatever Washington may say.

The Department of the Interior statement is carefully worded to avoid interference with states' rights to set independent wildlife policies. In my text reference is to "tax-supported wild life agencies," which of course refers to state and local controls, and not only or not especially to the federal agencies, whose policies are so often overruled locally. For example, although the Bureau explicitly mentions "lions" (pumas) as worthy of some protection, the State of Arizona has a perpetual open season on them, with no "justification" required, and it still pays bounties for their slaughter. Besides "control" and "removal" another euphemism for killing wild animals is "harvesting," and Arizona's beautiful pumas are rapidly being "harvested."

I do not doubt and I am thankful that within bureaus and agencies of the Department of the Interior there is honest sentiment in favor of the preservation, at least as nonextinction, of wildlife. I hope that some slow progress is being made in the face of strong lobbies and other political forces opposed to that sentiment. However, I do not believe that the critical statement made in the text of this chapter can be contradicted at present.

My reference to the use of herbicides in Vietnam also produces a second thought but not a contradiction. Early in 1968 the Department of Defense released a summary of a report on that use of herbicides, a report ordered and paid for by it at a cost of $62,000. Information from Vietnam itself being insufficient, the report was based primarily on observations elsewhere and, one can only infer, on subjective generalities. The summary makes clear that information over-all is quite inadequate, but it minimizes possible adverse effects and concludes that the use of herbicides in Vietnam, sufficient to treat 965,000 acres prior to preparation of the report, may actually have been beneficial. There seems to be no statement as to why a beneficial treatment is used as a weapon of offense or as to whether the poverty-stricken Vietnamese consider themselves benefited by massive destruction of their crops.

As I now write, there is some still quite dim chance of peace in Vietnam. If and when that comes to pass, we will cease defoliation and other offensive action there, but the morality of actions now already taken will not be changed.

CHAPTER FIVE

A first draft of this chapter was prepared as a contribution to a symposium on "The Impact of Education, Science, and Technology on the South," held at Duke University in April 1965. I was unable to attend, but my manuscript was read there and then was published in *The South Atlantic Quarterly,* Vol. LXIV, No. 4 (1965), pp.

478-495. Some of my remarks, especially those on the evils of tobacco, were directed with some unkindness at a local audience, but their subjects are of national significance and I have not expunged them here.

My adverse reference to Barzun, whom I admire as highly cultured on one side (which is a half more than most of us), is largely based on his book *Science: The Glorious Entertainment* (New York: Harper & Row, 1964). I explained my objections in "Barzun: the Glorious Entertainer," *Science*, Vol. CXLIX, No. 3614 (1964), pp. 38-39.

The fiscal 1966 figures for R & D are of course now passé, but current figures are not yet firm, and it is impossible to judge what proportions are, or will be, for uncommitted research, for committed or mission-directed research, and for development. The budget request for R & D in fiscal 1969 is $17.8 billions, including building as well as R & D in the literal sense. As I now write, demands for budget cuts related to proposals for a tax increase make it virtually certain that the authorized budget will be much less than the request. Even if the whole request were granted, the apparent increase over fiscal 1966 would be far below that necessary to keep up with "increasing costs" (another euphemism; it means inflation). In other words, the *request* (not likely to be granted) represents a sharp cut in real support for R & D. Moreover, in the proposed budget for fiscal 1969 a larger percentage of R & D funds will go to the Defense Department and the Atomic Energy Commission. Both will direct some of those increases into academic research, but that will be almost entirely for mission-directed research, and much of it for "missions" of which large numbers of loyal Americans do not approve. Prospects for government support of free research, which seemed fairly bright only a couple of years ago, are now being dimmed. It is even more true now than when I wrote these words three years ago that "expenditure for science in the United States is practically insignificant from a direct economic point of view. . . ." It does not particularly change the picture that expenditure for space, especially through NASA, is decreasing, although still enormous.

It is always tempting to produce more examples of the irrationality and antiscientific nature of advertising. More examples are not really needed, and their accumulation is perhaps too easy to be sporting. However, I find one quite irresistible because it manages to be profoundly misleading by telling an absolute, incontrovertible truth: "Don't kid yourself! Your cigarette isn't lowest in tars unless it is lower than C——."

My necessarily brief comment on firearm legislation is too mild, but more cannot be done without imbalance in this context. For a reasonably full story of the fanatical irrationality of gun-lovers and the dishonest and antisocial activities of the National Rifle Association see:

Harris, R. "Annals of Legislation: If You Love Your Guns." *The New Yorker* (20 April 1968), pp. 56-155.

Since I first wrote this essay, there has of course been an enormous amount of further discussion on extraterrestrial life. None of this has changed the situation in any way whatever. As mentioned before, NASA is now undergoing some retraction, but it is still a major element in the American scientific scene and in the (to me) odd pursuit of exobiology. Because the trend is now toward a better balance, I have deleted one paragraph about what I earlier considered undesirable trends in NASA.

The mention of D. K. Price refers to his fine book *The Scientific Estate* (Cambridge: Belknap Press, 1965).

CHAPTER SIX

After presenting varying lectures on this topic at Randolph Macon College, the University of Paris, and the University of Washington, I wrote this version as an essay. It was published in *Science*, Vol. CLII, No. 3721 (1966), pp. 172-178, and is here reprinted with slight editorial emendations.

The following citations include authorities mentioned in the text and also a few others useful in consideration of the various topics of this chapter. References on language, omitted here, are given for Chapter Eight.

Campbell, B. *Human Evolution* (Chicago: Aldine Publishing Co., 1966).

Clark, W. E. Le Gros. *Fossil Evidence for Human Evolution.* 2d ed. (Chicago: University of Chicago Press, 1964).

Darwin, C. *On the Origin of Species by Means of Natural Selection, or the Preservation of Favoured Races in the Struggle for Life.* (London: Murray, 1859).

——. *The Descent of Man and Selection in Relation to Sex* (London: Murray, 1871).

——. *The Expression of the Emotions in Man and Animals* (London: Murray, 1872).

DeVore, I., Ed. *Primate Behavior* (New York: Holt, Rinehart, & Winston, 1965).

DeVore, P. L., Ed. *The Origin of Man* (New York: Wenner-Gren Foundation, 1965).

Dobzhansky, T. *Mankind Evolving* (New Haven: Yale University Press, 1962).

Huxley, T. H. *Evidence as to Man's Place in Nature* (London: Williams & Norgate, 1863).

Leakey, L. S. B., P. V. Tobias, M. D. Leakey, and J. R. Napier. "The Olduvai Bed I Hominine with Special Reference to Its Cranial Capacity. Recent Discoveries of Fossil Hominids in Tanganyika at Olduvai and near Lake Natron. A New Species of the Genus *Homo* from Olduvai Gorge." *Nature*, Vol. CCII (1964), pp. 3-9.

Simons, E. L. "The Earliest Apes." *Scientific American*, Vol. CCXVII (1967), pp. 28-35.

Simpson, G. G. *Principles of Animal Taxonomy* (New York: Columbia University Press, 1961).

Washburn, S. L., Ed. *Classification and Human Evolution* (Chicago: Aldine Publishing Co., 1963).

CHAPTER SEVEN

This and the following chapter are in part reprises and paraphrases, in part glosses and expansions of two topics already discussed in a broader context in Chapter Six: race and language. Neither chapter has been previously published.

The general biological concept of species, subspecies, and races is sufficiently and excellently discussed in the previously cited book by Ernst Mayr. The application of those concepts to mankind is sensibly and excellently discussed in the book by Theodosius Dobzhansky also cited earlier. Brief statements by me, of which this chapter is in a way an expansion and application, appeared as "Races in Animal Evolution" and "Polytypism, Monotypism, and Polymorphism" in *International Social Sci. Journ.*, Vol. XVII (1965), pp. 148-153.

An anthropological account of present human races is:

Coon, C. S., and E. E. Hunt, Jr. *The Living Races of Man* (New York: Alfred A. Knopf, 1965).

It need hardly be said that some of his colleagues do not agree with Coon's interpretations, but they merit careful consideration and the descriptive data in this book are invaluable. There is also a good bibliography.

A book now dated nevertheless still merits citation and reading because it had a major effect on changing concepts of the human races:

Boyd, W. *Genetics and the Races of Man* (Boston: Little, Brown, 1950).

CHAPTER EIGHT

Like Chapter Seven (as noted above), this chapter is a reprise and gloss on a topic introduced in the broader context of Chapter Six. There is more repetition here, but I think it may be welcomed as giving clarity, emphasis, and a somewhat different orientation.

The psychologist quoted in the first paragraph is Leonard Carmichael. The phrase quoted is in, indeed it is the first sentence of:

Lenneberg, E. H., Ed. *New Directions in the Study of Language* (Cambridge: M.I.T. Press, 1964).

The other six chapters of that symposium are also highly relevant and useful for the present topic.

Whorf's ideas, cited now more for historical value than for present importance, are available in:

Whorf, B. L. *Language, Thought and Reality* (Cambridge: M.I.T. Press, 1956).

A brief but fully documented history (through the nineteenth century) of attempts to determine the origin of language is given by O. H. Marx in an appendix (pp. 443-469) to Lenneberg's book cited below.

The statement quoted from Jane Lancaster is in:

DeVore, P. L., Ed. *Origin of Man* (New York: Wenner-Gren Foundation, 1965).

That symposium also includes important papers on language by A. R. Diebold, Jr., D. Slobin, and T. A. Sebeok, with discussion by several other authorities.

The mention of Cassirer refers to:

Cassirer, E. *An Essay on Man* (New Haven: Yale University Press, 1944).

Although I have gladly credited Aristotle with not always being wrong, it happens that this very subject also illustrates the fact that he was rarely quite right. He recognized that only men have true language, but he thought that one of the distinctions is that language is produced mainly in the mouth, but animal sounds only by movement of air through the windpipe.

On writing and alphabets the most complete work, even though not now quite up-to-date, is:

Diringer, D. *The Alphabet* (New York: Philosophical Library, 1948).

On ethics, including C. H. Waddington's views, see Chapter Ten of this book.

The anecdote about the myna bird is from:

Mann, W. M. *Anthill Odyssey* (Boston: Little, Brown, 1948).

I am not here concerned with recent developments in formal analytical linguistics, a subject that has become so abstruse that only members of a small "in" group are likely to grasp it. The following are a few pertinent and interesting samples, far from exhaustive, of other studies relevant to this chapter:

Bryan, A. L. "The Essential Morphological Basis for Human Culture." *Current Anthropology,* Vol. IV (1963), pp. 297-301.

Diamond, A. S. *The History and Origin of Language* (New York: Philosophical Library, 1959).

Du Brul, E. L. *Evolution of the Speech Apparatus* (Springfield: Charles C. Thomas, 1958).

Fink, B. R. "Larynx and Speech as Determinants in the Evolution of Man." *Perspective in Biol. and Med.*, Vol. VII, No. 1 (1963), pp. 85-93.

Haldane, J. B. S. "Animal Communication and the Origin of Human Language." *Sci. Progress* (July 1955), pp. 385-401.

Haldane, J. B. S., and H. Spurway. "A Statistical Analysis of Communication in 'Apis mellifera' and a Comparison of Communication in Other Animals." *Insectes Sociaux*, Vol. 1, No. 3 (1954), pp. 247-283.

Kortlandt, A. [Remarks on Bryan's paper, cited above.] *Current Anthropology*, Vol. VI (1965), pp. 320-326.

Lenneberg, E. H. *Biological Foundations of Language* (New York: John Wiley & Sons, 1967).

Sebeok, T. A. "Animal Communication." *Science*, Vol. CXLVII, No. 3661 (1965), pp. 1006-1014.

——. "Zoosemiotics: Juncture of Semiotics and the Biological Study of Behavior." *Science*, Vol. CXLVII, No. 3657 (1965), pp. 492-493.

Spuhler, J. N., Ed. *The Evolution of Man's Capacity for Culture* (Detroit: Wayne State University Press, 1959).

CHAPTER NINE

I have lectured on this subject to a number of audiences, and a chapter on it is included in an earlier book (*This View of Life*, previously cited). The present treatment is new, however, and does not closely follow any prior lecture or publication. It was already drafted when the following highly relevant material appeared, and it was thereafter revised:

Neal, J. V., Chairman. "Symposium on Genetic Implications of Demographic Trends." *Proc. Nat. Acad. Sci.*, Vol. LIX (1968), pp. 649-699.

That symposium includes contributions by F. Seitz, J. F. Crow, D. Kirk, W. J. Schull (with three Japanese colleagues), J. V. Neal with N. A. Chagnon, and W. F. Bodmer. The paper by Crow (pp. 655-661) is particularly important for the present topic.

Having been accused of inventing the prediction that time will stop at a fixed date, I reluctantly give the authority for it:

Meyer, F. *Problématique de l'Evolution* (Paris: Presses Universitaires de France, 1954).

Teilhard's views and the reasons why they are scientifically untenable were sufficiently discussed in *This View of Life* (Chapter Eleven).

The idea that age of parents at time of childbearing affects the eventual number of their descendants is novel to some readers. Further explanation would have interrupted the chapter text, but is briefly supplied here. To use an extreme but possible hypothetical example, suppose a couple A completes a family of three offspring by parental age 20 and that the offspring in turn each complete families of three by age 20. At age 40 for couple A they will then have nine grandchildren. Their children and themselves will probably also be alive, so that there will be 14 living members of the family, counting A and their descendants. If another couple B, born at the same time as A, completes a family of three children at parental age 40, then at the given time (forty years after A and B were born), the B's will have three children and no grandchildren. The B's and their descendants will then number only five as against 14 for the A's, even though the A's and B's had equal numbers of children. The effect will of course increase even more rapidly as more generations pass.

Muller's often-repeated views on eugenics are included in the following symposium:

"Evolution and Man's Progress." *Daedalus*, Vol. XC (1961), pp. 411-586.

Muller's paper (pp. 432-450) is followed by extensive printed discussion (pp. 451-476), much of it (including mine) somewhat critical of his interpretations and proposals. The symposium also includes another important contribution by Crow (pp. 416-431), and much else relevant to the present chapter.

Dobzhansky's *Mankind Evolving*, cited previously, is again very pertinent here. The following are a few of the many other discussions of one aspect or another of this vast subject:

Dobzhansky, T. "Changing Man." *Science*, Vol. CLV, No. 3761 (1967), pp. 409-415.

Dubos, R. *Man Adapting* (New Haven: Yale University Press, 1965).

Sonneborn, T. M., Ed. *The Control of Human Heredity and Evolution* (New York: Macmillan, 1965).

Wolstenholme, G., Ed. *Man and His Future* (Boston: Little, Brown, 1963).

CHAPTER TEN

The first form of this chapter was the invited address at the annual meeting of the American Psychological Association, Chicago, September 1965. It was printed under the title "Naturalistic Ethics and the Social Sciences," in the *American Psychologist*, Vol. XXI (1966), pp. 27-36. (Reprints have incorrect volume, number, and date.) The present version has been somewhat modified.

It does not seem necessary to give citations for Locke and a few other classics. References to later authors cited or quoted in the text, plus a few others particularly relevant for anyone who wants to pursue this subject further, are given below. Brief comments on some of these are first made.

The concept of the supraorganism (or epiorganism) and the analogy between society and organism have been almost entirely replaced by studies of integration in societies that do not involve false and dangerous aspects of that concept and the analogy. Gerard and the Redfield symposium (in which some authors did not subscribe to the analogy) are cited for historical reasons only. Those views, most extensive in the 1930's, have waned since Hitler's war.

The book by Quillian is an excellent review of some phases in the search for naturalistic ethics. It also ably refutes such views as that the fact of evolution involves an unconditional *ethical* obligation to survive. However, when Quillian wrote, most students of evolution had abandoned, or indeed had never held, the views that he ascribed to them. His opposition to naturalistic ethics *in toto*, and not simply to certain obsolete forms of them, lacks force.

Waddington's interesting book includes one passage so strangely mistaken that I cannot cite the book for general readers without specific comment on this part of it. He says that "Dobzhansky and Simpson may be taken to suggest [the view] that a rational criterion for ethics cannot be acceptable because it limits the essential human faculty for the exercise of freedom. There is, however, still another way in which Dobzhansky and Simpson's words can be taken." He then goes on to say that there is also a nonrational approach, one that "does not lie within the sphere of conceptual thought," and that is exemplified by Taoism and Zen Buddhism. He concludes, "It seems rather likely that . . . one of the notions in their [Dobzhansky and Simpson's] mind was allied to this concept."

In such matters I do often agree with Dobzhansky, although it is not quite true that we have only one mind between us. (We have not even published anything jointly.) We do occasionally disagree, and I neither need nor venture to speak for Dobzhansky. For my part, I simply cannot imagine how Waddington can have reached those impressions if he has read anything that I ever wrote on the subject. Surely I need not again expound at length, but I do insist that it is precisely the "exercise of freedom," in that we can and must make choices and try to foresee outcomes, that makes it both possible and necessary to seek a rational criterion for ethics. I have no use whatever for a nonrational approach or for any system of values that does not lie within the sphere of conceptual thought. I believe that Tao and Zen, at best softheaded when not plain silly, are about as bad guides to desirable ethics as is Christian mysticism.

Most of the following works contain extensive bibliographies, leading still further into the tremendous literature on this subject.

Dobzhansky, T. *The Biological Basis of Human Freedom* (New York: Columbia University Press, 1956).

——. *Mankind Evolving* (New Haven: Yale University Press, 1962).

——. *The Biology of Ultimate Concern* (New York: New American Library, 1967).

Emerson, A. E. "Dynamic Homeostasis: A Unifying Principle in Organic, Social, and Ethical Evolution." *Sci. Monthly*, Vol. LXXVIII (1954), pp. 67-85.

——. "Human Cultural Evolution and Its Relation to Organic Evolution of Insect Societies." In *Social Change in Underdeveloped Areas: A Reinterpretation of Evolutionary Theory*, R. W. Mark, H. R. Barrington, and G. I. Blanksten, Eds. (Cambridge: Schenkman, 1965), pp. 50-67.

Freud, S. *New Introductory Lectures on Psycho-analysis* (New York: W. W. Norton, 1933).

Gerard, R. W. "A Biological Basis for Ethics." *Philos. of Sci.*, Vol. IX (1942), pp. 92-120.

Huxley, T. H., and J. S. Huxley. *Touchstone for Ethics* (New York: Harper, 1947).

Kinsey, A. C., W. B. Pomeroy, and C. E. Martin. *Sexual Behavior in the Human Male* (Philadelphia: W. B. Saunders Co., 1948).

Leake, C. D. "Ethicogenesis." *Sci. Monthly,* Vol. LX (1945), pp. 245-253.

Lerner, I. M. *Genetic Homeostasis* (New York: John Wiley & Sons, 1954).

Lotka, A. J. "The Law of Evolution as a Maximal Principle." *Human Biol.,* Vol. XVII (1945), pp. 167-194.

Patten, B. C. "An Introduction to the Cybernetics of the Ecosystem: The Trophic-dynamic Aspect." *Ecology,* Vol. XL (1959), pp. 221-231.

Quillian, W. F., Jr. *The Moral Theory of Evolutionary Naturalism* (New Haven: Yale University Press, 1945).

Redfield, R., Ed. *Levels of Integration in Biological and Social Systems.* (Lancaster: Cattell, 1942).

Shakow, D. "Ethics for A Scientific Age: Some Moral Aspects of Psychoanalysis." *Psychoanalytic Rev.,* Vol. LII, No. 3 (1965), pp. 335-348.

Simpson, G. G. *The Meaning of Evolution.* Rev. Ed. (New Haven: Yale University Press, 1967).

——. *This View of Life* (New York: Harcourt, Brace & World, 1964).

Waddington, C. H. *The Ethical Animal* (London: Allen & Unwin, 1960).

Zipf, G. K. *Human Behavior and the Principle of Least Effort* (Reading: Addison-Wesley, 1949).

Index